SURVEY RESEARCH FOR GEOGRAPHERS

Ira M. Sheskin
Department of Geography
University of Miami
Coral Gables
Florida

RESOURCE PUBLICATIONS
IN GEOGRAPHY

Library of Congress Card Number 84-73382
ISBN 0-89291-187-5

Library of Congress Cataloging in Publication Data

Sheskin, Ira M., 1950-
    Survey Research for Geographers.

    (Resource publications in geography)
    Bibliography: p.
    1. Geography — Methodology. 2. Social surveys.
I. Title. II. Series.
G70.S52 1985   910'.723   84-73382
ISBN 0-89291-187-5

Publication Supported by the A.A.G.

Graphic Design by CGK

Printed by Commercial Printing Inc.
State College, Pennsylvania

Foreword

Survey research has become an important methodology for geographers, primarily in our role as social scientists. We have used surveys to assess people's views of natural hazards, to document travel behavior, to ascertain locational preferences, and to understand spatial knowledge. Unfortunately, we have not always been aware of the difficulty of undertaking survey research, and in some cases, we may have inadvertently violated basic ethical principles. Failing to recognize logistical and substantive challenges in survey research, substantial time and money may have been invested to produce questionable results.

This monograph has been prepared to guide geographers in the use of survey research. Although it is not the first such document, it is perhaps the most comprehensive reference within the profession that can be used as a set of principles guiding us in the development of a survey research project. By drawing upon a number of surveys done in geographical research as well as upon his own experience with a variety of survey techniques, Ira Sheskin provides a number of examples to illustrate the basic issues in survey research and to point out some of the 'art' as well as 'science' in using this methodology.

The **Resource Publications in Geography** are sponsored by the Association of American Geographers, a professional organization whose purpose is to advance studies in geography and to encourage the application of geographic research in education, government, and business. Tracing its origins to the AAG's *Resource Papers* (1968-1974), the **Resource Publications** continue a tradition of presenting geographers' views on timely public and professional issues to colleagues, students, and fellow citizens. Views expressed, of course, are the author's and do not imply AAG endorsement.

The author, editor, and Advisory Board trust that this book will enhance the quality of survey research undertaken by geographers. We also hope that it will stimulate greater attention to research methodologies in human geography. Most fervently, we anticipate that this volume will give additional impetus to the development of a professional standard of ethics in geography by calling attention to the ethical issues involved in survey research methods.

C. Gregory Knight, *The Pennsylvania State University*
Editor, Resource Publications in Geography

Preface and Acknowledgements

Geographers make significant use of survey research techniques. Survey results are reported in numerous journal articles, paper presentations, theses, and dissertations within geography and cognate disciplines. They also are discussed on an almost daily basis in the popular press. Thus, it behooves both professional and student geographers to expand their knowledge of survey research.

One common thread, which runs through most of the geographic literature employing survey research methods, is the concentration upon the reporting of survey results rather than survey methods. This may be viewed as proper. When geographers began to use quantitative methods, many articles were published in which the focus became the method, rather than the geography; the geographic example was employed only to help explain the method. Yet, with respect to survey research, the opposite has occurred. An extensive examination of the past ten volumes of *The Professional Geographer, The Annals of the Association of American Geographers,* and *Economic Geography,* and the past five volumes of *Urban Geography,* reveals very little discussion of the methodology of surveys whose results are being presented. This is unfortunate for two reasons. First, the validity of the results is difficult to judge when response rates, sampling frames, question wording, and other topics are unspecified. Second, discussion of the technique within the literature would assist other researchers. Thus, the use of survey research techniques by geographers is widespread, but discussion of these techniques in the geographic literature is limited. This *Resource Publication* attempts to ameliorate this situation: it should serve as a reference for conducting surveys for both professional and student geographers and also as a guide to the survey research literature.

Within geography, instructors of field and quantitative methods courses should find this volume to be of significant utility. Many geographic field courses cover some aspects of survey research, as evidenced by the existence of chapters on survey research in recent field methods textbooks. On the other hand, a perusal of any of the textbooks on quantitative methods in geography (as well as statistics books used in other disciplines) indicates that such courses generally assume that the data to be subjected to t-tests and regression analyses are available from secondary sources. At a minimum, these books assume that the reader knows the basics of survey research and therefore limit discussion to sample size and sampling methods. Whereas survey research is not in and of itself a 'quantitative method,' it often provides the data used in quantitative methods courses. A superior understanding of quantitative methods will derive from a situation in which the student has defined a problem, collected supporting data, and then applied the appropriate statistical techniques. At that point, the data for a *chi-square* problem begin to gain significance within the context of a research problem, and cease to be just a column of numbers to be punched into a calculator. Particularly for human geographers, survey research is a primary tool for data collection. Thus, those who gain an appreciation of survey research will be able to attain a superior grasp of quantitative data analysis methods.

The author would like to acknowledge the many fine suggestions of the anonymous reviewer and the comments offered by Peter O. Muller on an earlier draft of the manuscript. Most of all, thanks are due to my wife, Karen, who spent innumerable hours both editing the manuscript and making a number of important substantive contributions.

Ira M. Sheskin

Contents

List of Figures

List of Tables

1

To Survey or Not to Survey . . .

It might seem paradoxical to introduce a volume encouraging the use of survey research techniques by geographers with a chapter implying that the use of surveys should be limited. Yet, many survey efforts are unnecessary, much of the data collected are never utilized, and some survey methodologies are so flawed that unwarranted conclusions are reported. Recently, a panel of the Committee on National Statistics of the National Research Council's Assembly of Behavioral and Social Sciences (Heggemeier 1982) recommended that the time has arrived to:

> . . . take surveys and polls seriously. Just as one would not endorse the idea that anyone can play doctor, administering any and every drug, one should not endorse the concept that anyone can do or interpret a poll or that anyone's poll is necessarily as good as anyone else's . . . resources should be husbanded so that fewer surveys and polls could provide more and better information . . . too many surveys of mediocre and inferior quality [have been produced] that do not accomplish anything useful. . . . There are others that should be exploited more fully: too much time and effort is spent in collecting survey results and too little finding out what responses mean. Still other surveys are based on population samples too small to support conclusions.

Granting agencies (particularly the National Science Foundation) may contribute to the proliferation of surveys. Academic researchers are often judged by the size of their grants. Grants often propose costly collection of new data more often than further analysis of existing data.

A number of additional factors contribute to survey proliferation. First, numerous United States federal government regulations require proof of 'need' before federal monies are disbursed to local governments. For example, federal regulations require that local governments provide evidence that their transit system is serving all persons, regardless of race or income, before federal monies can be used to purchase transit vehicles. Such evidence is usually collected via survey research.

Second, the popular press devotes significant space to survey findings. A number of newspapers employ a survey researcher in a full-time capacity. According to the new 'survey editor' of the *Miami Herald* (1983):

> We have been devoting more news space to our own polls and to those done by others. Academic social scientists see survey results as a vehicle for gaining some notoriety for themselves and their institution.

Third, the results of many surveys are just plain interesting, providing information for which Americans seem to have an insatiable appetite (Norback 1980). Each year at the EPCOT Center at Florida's Walt Disney World, thousands of persons volunteer their opinions on current issues and enjoy seeing their answers, combined with those of others' in the audience, displayed immediately on a large screen. Fourth, although surveys are costly, recent refinements of telephone survey techniques have significantly reduced costs. Finally, surveys have become important tools for marketing research and political decision making.

Survey research techniques have not received the attention in geography that they have in such fields as sociology, psychology, and marketing. Thus, it is possible that the level of misuse within geography is greater than in the other social sciences. Yet, survey research is a long-established method of geographic field research which has:

> . . . an ancient and honorable tradition in geography. . . . As geography began to mature its practitioners realized that knowledge and insight into the nature of places and areas must be based on more careful observation, more precise measurement, and more accurate recording of the features in which they were interested (Lounsbury and Aldrich 1979: ix).

Particularly when such "features" include the behavioral characteristics of human subjects, survey research becomes a primary data collection tool. Surveys are probably not what some geographers envision when urging others in the discipline to 'get their boots dirty in the field' (particularly when the selected survey mechanism is the telephone). Nevertheless, survey research clearly fits within the realm of geographic fieldwork, and its inclusion in a field course is highly appropriate. Although not every significant geographical research problem requires the collection of original data via survey research, appropriate research design often relies upon survey data that have been collected by others. The inclusion of chapters on survey research in two recent textbooks on field techniques in geography (Lounsbury and Aldrich 1979; Stoddard 1982) underscores the continuing acceptance of survey research as a field tool.

Survey Research and Scientific Explanation

There are two routes to scientific explanation: *induction* and *deduction*. Induction proceeds from particular cases to universal statements; deduction proceeds from some general *a priori* premise to statements about particular events (Figure 1). In practice, induction implies that one collects and orders data and then develops laws and theories. Deduction implies that one has hypotheses concerning the ultimate form of laws and theories and then proceeds to collect data that will facilitate verification of such hypotheses. While this distinction (Harvey 1969:32-6) is a useful one, most scientific efforts involve a combination of these modes of reasoning.

Survey research contributes to both induction and deduction during the "measurement" stage shown in Figure 1. Put simply, if one is designing a questionnaire as part of an inductive research design, then questions are defined with no reference to the statistical relationships that ultimately will be examined. If a deductive approach is being employed, then one first states a series of hypotheses and then defines questions that will specifically facilitate the verification of these hypotheses. Unlike other research methodologies (for example, data collection from maps or aerial photographs), if a critical variable is not included in the original survey, it is rarely possible to resurvey the sample.

In practice, most researchers design questionnaires in a deductive fashion with inductive overtones. That is, a series of hypotheses or 'problems' are defined, and questions are formulated to address these issues in a deductive manner. Most surveys also include several questions designed to gather information whose relationship to the overall purpose of the survey may not be obvious. In such cases, the data are being collected inductively and a testable hypothesis is formulated *a posteriori.*

There is an inverse relationship between the length of a questionnaire and the proportion of subjects who will respond (the response rate). Thus, the art of survey

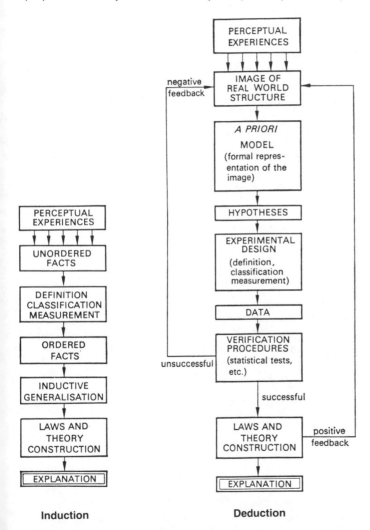

Induction   **Deduction**

FIGURE 1   ROUTES TO SCIENTIFIC EXPLANATION. Reproduced from Harvey (1969:34) by permission of Edward Arnold (Publishers) Ltd.

research is to ensure the inclusion of all questions relating directly to the deductive process and to limit those of an inductive nature. Hopefully, responding to the questionnaire is not an arduous task.

As an example of the inductive/deductive process, the *Population Study of the Greater Miami Jewish Community* (the GMJF Study; Sheskin 1982) had three purposes: (1) a demographic profile of the Jewish population; (2) a survey of the need for services; and (3) a survey of the willingness of the population to support services. A ten-minute telephone and a 12-page mail survey were designed to meet these a priori aims. Among included questions of an *inductive* nature were "Do you own or rent your residence?" and "Do you live in a single-family home, a condominium, or an apartment?" While the consultant had formulated some a priori hypotheses which would make these questions important, the research sponsor (the Greater Miami Jewish Federation) could see no pertinent reason for them and requested their exclusion to shorten the questionnaire. After much discussion, it was agreed that these questions would be included if they did not cause the questionnaire to expand by a page. These questions proved extremely useful as concerns arose about neighborhood stability and the possibility of providing on-site services to the elderly living in condominium housing. The two questions were originally derived in an inductive fashion, in the sense that their relationship to the stated purposes of the survey was not initially obvious. They later provided critical data for some a posteriori hypotheses, developed after the responses were classified and tabulated.

A second example is provided by the *University of Miami Travel and Parking Survey* (Sheskin and Warburton 1983). No a priori deductive hypothesis existed to justify asking respondents to identify their gender on this self-administered questionnaire. Yet the result proved intriguing: intent to use the new Dade County METRORAIL system on a daily basis was five times higher for males than for females.

A well-designed survey should be centered around clearly-defined hypotheses or problems. If a few 'stab-in-the-dark' questions are included, the overall integrity of the survey procedure will not be compromised. If, on the other hand, an entire survey is designed in an inductive fashion, the response rate suffers because respondents sense that the questions have no real focus. In addition, much of the data generated in such a fashion may not be used. Thus, surveys should have clearly-defined purposes and should consist, for the most part, of questions related directly to these purposes.

A Brief History of Survey Research

One of the first attempts at survey research in recorded history was divinely inspired, when God spoke unto Moses in the wilderness of Sinai, saying:

> Take ye the sum of all the congregation of the children of Israel, by their families, by their fathers' houses, according to the number of names, every male, by their polls; from twenty years old and upward (Numbers: 1:2).

Moses knew nothing of the concept of random sampling or the stratification of results, yet the need for such data collection (in this case, for conscripting an army) was evident. Both the ancient Egyptians and Romans used periodic censuses as a basis for taxes and military conscription.

In the 1770's, British reformer John Howard used survey research to study the effects of prison conditions on the health of the inmates in various European countries.

Some years later, Frederic LePlay used surveys in a study of income and expenditures of European families. An early use of surveys for political purposes occurred in 1880, when Karl Marx mailed questionnaires to 25,000 French workers to examine employee exploitation. Among the questions asked were:

1) Does your employer or his representative resort to trickery in order to defraud you a part of your earnings? and, 2) If you are paid piece rates, is the quality of the article made a pretext for fraudulent deductions from your wages? (Bottomore and Rubel 1956:208).

In 1886, English statistician Charles Booth conducted a massive study of poverty using survey research. All four of these early survey researchers experienced difficulty maintaining scientific objectivity in their zest to solve social problems, a dilemma which continues to plague survey research to this day.

Survey research may have its roots in the distant past, but the sample survey as we know it today is, for the most part, a product of twentieth century American science (Babbie 1973). Even as recently as 1930, the *Encyclopedia of the Social Sciences* warned that sampling human populations is so difficult that complete enumeration should be preferred. Major contributors to the development of survey research have included:

(1) Rensis Likert and his associates at the United States Department of Agriculture, who made major advances in applying area probability sampling methods to studies of attitude and behavior (Warwick and Lininger 1975:3);

(2) The United States Bureau of the Census, which conducts numerous sample surveys in addition to the constitutionally-prescribed decennial enumeration of the entire population;

(3) Commercial pollsters or polling corporations such as Gallup, Harris, and Roper;

(4) Social researchers such as Samuel A. Stouffer and Paul F. Lazarsfeld (Babbie 1973:92-93);

(5) Survey research centers at academic institutions, including the National Opinion Research Center at the University of Chicago, the Institute for Social Research at the University of Michigan, and the Survey Research Center at the University of California, Berkeley; and,

(6) Professional organizations such as the American Association for Public Opinion Research, which publishes the leading journal on survey research methodology, *Public Opinion Quarterly*.

As will be seen below, two technological inventions — the telephone and the computer — have also played a major role in the advancement of survey research.

Survey Research in Geography

Geography and geographers have not played a major role in the development of the modern sample survey. However, the use of survey research techniques by geographers is widespread. Many journal articles and papers presented at professional geography meetings report survey results.

Although survey research has been used extensively in geography, discussion of the technique and its application to spatial problems has not been adequately presented in the literature. Recent textbooks in field methods (Lounsbury and Aldrich 1979; Stoddard 1982) devoted chapters to the topic, and Dixon and Leach (1978a, 1978b, 1984) presented significant overviews. The paucity of work is best illustrated by the fact that Dixon and Leach cite only one work from the geographic literature in their bibliography (Walker 1976 — a sociologist writing in the British journal *Area*). Moreover, in Lounsbury and Aldrich's introduction to survey research, only two dated works are cited (Highsmith 1962; Kniffen 1962). A bibliography prepared for the Council of Planning Librarians by Friberg (1974) cites only one additional source (Thomas 1971); I have found only three more references (Stimson and Ampt 1972; Clark and Unwin 1980; Edwards and Shaw 1982).

Some geographic subdisciplines have made substantial use of survey research. These include research on natural hazard perception, travel behavior, cognitive distance studies, industrial location, and residential search behavior. However, geographers appear to be using survey techniques without critical discussion in their professional journals and textbooks.

Burton, Kates, and White (1970) employed survey research in their investigation of natural hazards perception. Others followed their lead, including Kromm (1973), working on air pollution in Yugoslavia; Brooks (1973), studying drought in northeastern Brazil; Harvey, Frazier, and Matulionis (1979), examining airport noise in Buffalo; Baumann and Sims (1979), working on flood insurance; and Simpson-Housley and Curtis (1983), studying earthquake occurrence in New Zealand.

Geographers researching urban transportation, particularly travel behavior, have made extensive use of survey research. Examples include Ojo's (1973) study of agricultural work trips in Nigeria's Yorubaland; Lentnek, Lieber, and Sheskin's (1975) and Lentnek, Charnews, and Cotten's (1978) examinations of shopping trips in Mexico; Ackerman's (1975) study of shopping trips in Argentina; Halvorson's (1973) work examining the relationship between income and journey to work; Wheeler and Thomas' (1973) study of work trips in Honduras; Stutz's (1974) examination of social trips in San Diego; Kocher and Bell's (1977) work on ride-sharing and absenteeism; Schuler's (1979) study of store choice; Monroe and Halvorson's (1980) work on the impact of pricing policy on transit ridership in Chicago; and Timmermans' (1983) study of grocery shopping behavior.

A third research area employing survey research explores cognitive distance and learning behavior. Some examples include Golledge, Rivizzigno, and Spector (1973), concerning the learning process of new residents about their city; MacEachren's (1980) study of the perception of distance to stores; and Brown and Broadway's (1981) comparison of cognitive distance among students living in small and large places.

A fourth research cluster, exemplified by a number of related articles in *Economic Geography,* focuses on the locations of industrial firms through surveys of the firms' managers (Scott 1983, 1984). Gibson and Worden (1981) used an in-field and telephone survey to examine the economic base multiplier. Seley (1981) employed interviews in studying small businesses. Mail surveys were used by McConnell (1979), to analyze exporting behavior; Daniels (1982), to investigate office location behavior; and Wheeler (1981), to study the effect of geographic scale on manufacturing location.

A fifth research area employing survey techniques concerns residential search behavior and neighborhood development. Examples include Davis and Casetti's (1978) study of neighborhood preferences among blacks; Zonn's (1980) study of residential search behavior among blacks; Clark and Huff's (1980) work on residential mobility rates; Roseman and Williams' (1980) and Meyer's (1981) examinations of metropolitan to non-metropolitan migration; Mercer and Phillips' (1981) work on homeowners' decisions to rehabilitate their properties; and Cutter's (1982) study of suburban residential satisfaction.

Other examples of geographic work using survey research include Caruso and Palm's (1973) work on the perception of change in neighborhoods; Palm's (1973) study of social areas; Shannon and colleagues' (1975) work on accessibility of development; Dear's (1977) work on mental health; Taylor and Neville's (1981) study of managers of industrial firms in Singapore; and Joseph and Poyner's (1982) examination of public service utilization.

These research areas are representative of applied survey techniques in geography; others are cited in later chapters.

Alternatives to the Sample Survey

Even the simplest survey is considerably more complicated and time-consuming than one might assume. Therefore, once a research problem has been identified, alternatives should be fully explored before committing significant time and resources to collecting new data through a survey. Fowler (1984:11) has stated that, "unfortunately, some people think of a survey as a first effort to try to learn something about a population; more appropriately, a full-scale probability survey should be undertaken only after it is certain that the information cannot be obtained in other ways and the need for the information is significant."

At least thirteen alternatives to a sample survey can be discussed: using survey data collected by others, the U.S. Census Public Use Sample, 'rider' questions, a case study approach, participant observation, experiments, focus groups, physical traces, archives, observation, hardware methods, content analysis, and simulations.

First, survey data collected by others may be available. Many universities belong to the University of Michigan Data Consortium from which one can obtain data collected in numerous surveys. Many local planning agencies, especially transportation planning organizations, have collected survey data that they are willing to share. *The Gallup Opinion Index, Public Opinion Quarterly,* and *Public Opinion* are other sources for such data. A number of geographic studies were based upon the use of transportation data collected by a local planning agency. Recker and Schuler (1982) used such data in an analysis of travel behavior and urban form; Stutz (1974) and Stutz and Butts (1976) used data from a major origin-destination survey in San Diego to examine social and recreational trip making. Roseman and Knight (1975) used data from the National Opinion Research Center in studying the migration of blacks.

Second, national censuses are a rich source of data. Whereas most aggregated census data are reported for areas (tracts, counties, etc.), the U.S. Bureau of the Census does release records of a random sample of 5% of all households (the Public Use Sample), identified only by the census tract in which they are located (Boswell and Jones 1978). Samples are available of 1 in 100, 1 in 1000, and 1 in 10,000 households for political units of 250,000 population or more. Thus, census data could be utilized in a study which, for example, included a comparison of two ethnic groups with respect to mode of transportation to work.

Third, 'rider' questions can often be used to obtain answers to a limited number of questions. This procedure involves simply adding questions to a survey being conducted by another party. Sternstein (1974) used this technique effectively in a study of migration to and from Bangkok.

A fourth possible data collection technique involves the use of case studies. In a study of the locational decision-making process for suburban office buildings, for example, it may very well be that in-depth studies of a limited number of projects might suffice. Such case studies might help to define issues which could be used later in a survey.

Fifth, participant observation (Friedrichs and Ludtke 1975; Horvath 1970) is a technique often used by anthropologists and sociologists in which the researcher acts as a participant in the activity under study. Thus, instead of conducting a survey to ascertain the satisfaction of rail patrons with a new rail system, a researcher might 'ride the rails,' observing the manner in which riders use the system, listening to their comments, and initiating conversations. There are two major drawbacks of this technique. It is very expensive, requiring a major time investment on the part of a professional person. Experimenter contamination may occur with the presence of the professional influencing others' behavior and comments (a problem it shares with survey research). Examples of the use of this technique in geography include Cybriwsky's (1978) work on social aspects of neighborhoods, Ley's (1974) work on the inner city, Symanski's (1974) work on prostitution in Nevada, and Rowles' (1978) investigation of the spatial behavior of the elderly.

Sixth, 'experiments' (most often used by psychologists) involve a relatively small number of individuals (the treatment group) who are subjected to some pretreatment procedure (such as an interview or questionnaire) which collects data on various individual characteristics prior to their being subjected to some experimental procedure. Later, the subjects are administered a post-treatment procedure. Sometimes, to provide a means for comparison, a control group is also used, undergoing both sets of tests, but without the experimental procedure. An example of this type of research in geography is the work of Golledge and others at Ohio State University (King and Golledge 1978) who studied the manner in which new residents learn about the urban environment. Cartographers (Lloyd and Steinke 1977, for example) have employed this type of research to test various map designs. Winett and colleagues (1979) used a procedure in which a treatment group was provided with electric meters inside their homes to discern their effectiveness in promoting conservation. Marilyn Brown, a geographer at Oak Ridge National Laboratory, is currently designing an experiment in which a treatment group will be provided with energy conservation literature, with energy consumption measured before and after literature delivery. Changes in consumption may then be compared with a control group receiving no literature.

Seventh, 'focus groups' can be used to collect detailed information. It is often difficult in survey research to ask detailed questions requiring extensive answers. In a focus group, small groups are assembled to obtain such information. Often, group dynamics can lead to more considered responses than might be obtained in a written questionnaire, although these same group dynamics can sometimes suppress ideas. This methodology can be very fruitful, but it is important to recognize that it is unlikely that one could obtain the consent of a random sample of the general population to attend such a group. Also, the conditions under which the information is gathered might affect data validity, although this disadvantage is present in a number of the alternative methods as well. (This is also a major problem in survey research: The interviewer-interviewee relationship in survey research is a contrived environment which often creates data, rather than collecting it.) Therefore, the results of a focus group can rarely be generalized to the population as a whole.

An eighth method of collecting survey-type information without resorting to a survey, the use of physical traces, does not involve contacting human subjects. Such techniques have been called "unobtrusive measures" by Webb and colleagues (1966). Suppose a geographer was interested in studying the selection of different makes of automobiles by persons from different counties. Examining license plates yields information (in many areas) on county of origin. An observer in a parking lot or at an intersection could record both automobile type and county. As a second example, the level of erosion of various paths used by those visiting a park might provide important clues for a recreational geographer studying the levels of usage of various facilities. Also, some social scientists use domestic refuse as a source of data, as illustrated by a recent issue of the *American Behavioral Scientist* entitled "Household Refuse Analysis: Theory, Method, and Applications in Social Science" (Rathje and Ritenbaugh 1984).

Ninth, archives can often be a useful source of data (Norton 1984). A population geographer may use actuarial records to examine regional differences in such variables as birth out of wedlock and fertility levels. City directories or telephone directories may be used to trace the residential mobility of a population. Existing maps and/or aerial photographs may be consulted to yield relevant data for research.

Method ten involves the use of simple observation. In a study concerning the mode of access to METRORAIL stations in Dade County, Florida, two possibilities existed for data collection. First, a one-question questionnaire (How did you get to the METRORAIL station this morning?) could be designed. Second, the study could be accomplished by simple observation plus the station turnstile counts. A written questionnaire would probably suffer from enormous non-response, particularly because county law prohibits the production of a Spanish questionnaire and a significant proportion of the ridership is presumed to not read English. An oral questionnaire, administered as respondents passed through the turnstiles, would probably interfere with the operation of the system. The chosen method was to count the numbers of cars entering METRORAIL parking lots or dropping people off at the kiss-and-ride locations, and the number of persons transferring from bus to rail. Subtracting the total of these counts from the turnstile counts yielded the number of persons who walked to the station. While some inaccuracies were introduced by persons parking cars at some distance from the station and walking (to avoid the parking fee), these errors may very well be smaller than those introduced by the nonresponse bias of a questionnaire procedure. Another, but important, reason for this choice was that a more complete on-board survey was to be administered within the year, and there was a desire not to 'over-survey' the rider population. There is little justification for conducting a survey when an alternative methodology exists. A similar methodology was used by Ojo (1973) in a study of travel behavior to agricultural work in Southwestern Nigeria, except that every tenth person was also interviewed.

Eleventh, some studies have used hardware to collect data. These techniques include the use of audio and/or video tape recorders, traffic counters, cameras, devices which monitor television viewing, and remote sensing devices. For example, if an agricultural geographer was interested in obtaining information on the types of crops planted in an area, the use of remotely-sensed imagery (such as aerial photographs) might well yield such information with considerably less effort than employing a survey. Leinbach (1973) used a sample of telephone calls to examine spatial interaction in western Malaysia. While few ethical questions arise from the use of satellite images, hidden tape recorders or photographs may, in some cases, raise such questions.

A twelfth method of avoiding the collection of data via a sample survey is content analysis (Holsti 1969; Krippendorf 1980). This method involves analyzing textual material and counting the number of times certain key words or phrases are mentioned. For example, a political geographer may collect random samples of newspapers from various city types and sizes and compare the number of times various countries are mentioned in different cities. Some type of scoring method might be devised to rate whether or not places are mentioned in a positive or negative light. This may reveal information about the "international outlook" that exists in each city. In the same vein, an urban geographer might examine local small-town newspapers between major city A and major city B, looking for relative mention of either city A or B. Each small town would then be assigned to the relevant sphere of influence. While such techniques have not been used to any great degree in geography, Zelinsky (1980) has used content analysis to define America's "vernacular regions." As more textual material becomes available for microcomputers, counting the number of times places are mentioned should be made significantly easier.

A final alternative to a survey is a simulation study (Yeates 1974), in which a computerized mathematical model of the real world is used to predict how conditions might change in response to new values. Morrill's (1965) study of black ghetto expansion in Seattle was the seminal work in this area of geography. Role playing, in which persons act out roles in a mock confrontation, is another type of simulation. The Metfab game of the High School Geography Project is one such example (HSGP 1979).

Haggett and colleagues (1977) suggested that 95% of the work of geographers derives from archival sources, including maps and census data. Participant observation, controlled experiments, focus groups, and content analysis have seen little use in the discipline. Reviews of recent journals of the Association of American Geographers suggest that the past ten years have seen considerably greater use of survey research techniques.

Webb and others (1966:34) advanced a point of view, however, that geographers should heed as they become more involved in survey research techniques:

> It is only when we näively place faith in a single measure that the massive problems of social research vitiate the validity of our comparisons. We . . . argue strongly . . . for a conceptualization of method that demands multiple measurement of the same phenomenon . . . Over-reliance on questionnaires and interviews is dangerous because it does not give us enough points in conceptual space to triangulate.

Thus, survey research should be used only when no other technique could yield similar results. Nevertheless, survey research should rarely be one's *only* methodology. A strong case can be made for the use of survey research in situations where one wishes to infer the characteristics of a population from a sample. But one also can be made for using such methodologies as case studies, participant and simple observation, and the consultation of archival sources, prior to composing a questionnaire. Such techniques can be extremely helpful in defining the issues to be addressed in a survey instrument. Thus, not only will non-survey research techniques provide valuable insights in their own right, but they are also a first step in the questionnaire design process. For example, an intensive study of the reasons for the location of a small number of shopping centers might yield information which could then be incorporated in the design of a questionnaire of a random sample of all shopping centers.

This introductory chapter has argued that numerous methodologies for data collection exist, that survey research is somewhat unique in its ability to allow inference to a larger population, and that survey research should rarely be used as the only source of information when examining an hypothesis. However, when survey research is to be employed as a data collection method, it must be done correctly *the first time.* Unlike other less expensive methodologies, if a flaw is found in the procedures, it is rare that a correction can be implemented. If a truly random sample of the population is not interviewed, or if a critical question is omitted from the survey instrument or asked incorrectly, no easy remedy exists. Thus, it behooves the social scientist to become an "expert" (or engage an expert consultant) prior to implementing any survey.

Survey Research and Ethical Considerations

An overriding consideration in the decision to employ a sample survey is the ethics of the procedure. To a greater extent than for many of the alternative methodologies, survey research involves the social scientist in a forum in which interaction with the general public occurs in a manner that can only be viewed as an invasion of privacy, be it a personal interview at home or via telephone, the interception of an individual on the street, or, to a lesser extent, a mail survey. This is not to say that there are not significant ethical problems with some of the other techniques. For example, in participant observation, one such problem arises when the group or individual being observed is not informed that the social scientist is involved in research. Ethical problems often arise as well when human subjects are subjected to experimental conditions.

When a potential respondent is being asked to give time to answer what are sometimes sensitive questions about demographic characteristics, attitudes, behaviors, or beliefs, ethical considerations are important. A basic guideline is that the researcher should make certain that no individual suffers any adverse consequences as a result of a survey (Fowler 1984:135). While geographers are most often involved in surveys that do not concern particularly sensitive matters, such as sexual behavior, seemingly non-sensitive information may also cause a respondent to be harmed. In a one-question telephone survey I designed to estimate population size given the numbers of households (How many persons live in your household?), an elderly woman suffering from mental health problems became extremely upset by the survey (because she lived alone), to the point of necessitating a visit by a social worker. While such instances are rare, at times simply asking respondents about certain beliefs or attitudes causes discomfort, either because they are embarrassed, or because it causes them to confront some perceived 'failures,' such as a lack of religiosity, failure to

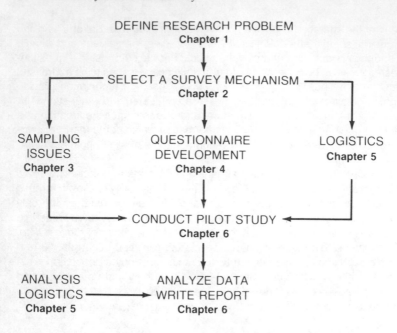

FIGURE 2   THE SURVEY RESEARCH PROCESS

visit cultural facilities, or a lack of tolerance for the feelings of others. Thus, one must always question whether the information to be gained from a survey is significant enough to justify asking people to undergo the survey process. Ethical considerations are revisited at various points throughout this volume and in the final chapter.

Overview of the Survey Research Process

Figure 2 illustrates the major steps involved in undertaking a sample survey of a human population. Following the definition of a research problem, the first task is to review the relevant literature and to examine the possibility of using one of the alternative procedures for data collection outlined above. If these procedures fail to yield appropriate data, then a sample survey is required (Figure 3).

Any survey consists of eight major steps:

(1) Selection of the survey mechanism (*e.g.*, telephone, mail);
(2) Selection of the sample;
(3) Development of the questionnaire;
(4) Organization of a series of logistical issues;
(5) Development of procedures for computerizing the results;
(6) Implementation of a pilot study;
(7) Implementation of the main survey; and
(8) Data analysis and report writing.

The order of the eight procedural steps is fairly important. Step 1 must occur first, although it might be possible to test alternative survey mechanisms in a pilot study, thus deferring this decision. Steps 2, 3, and 4 should take place concurrently and must be completed before pilot testing. The coordination of the project, so that everything is ready on 'survey day,' is crucial to a successful survey effort. Step 5 need not be completed prior to the pilot study. Upon completion of the pilot study, decisions made earlier may be revised as necessary. If such revisions are significant, it may be necessary to perform a second pilot study. If the changes are minor, one may then proceed with the main survey. The research is not complete until the data are analyzed and the report is written. Chapter 2 discusses the selection of the survey mechanism. The third chapter outlines the procedures necessary to select an appropriate sample of respondents. Chapter 4 addresses questionnaire development and Chapter 5 discusses survey logistics and procedures for computerizing the results. Finally, Chapter 6 discusses the conduct of the pilot and main surveys and the analysis of the data.

The Eight Example Surveys

This volume draws upon both the survey research literature and my own experience in designing a number of surveys, including:

The GMJF Demographic Study, a telephone/mail survey of the Jewish community of Dade County, whose principal purpose was to provide the necessary data to determine the types and locations of future capital facilities for the Jewish community of Dade County. The study was sponsored by the Greater Miami Jewish Federation (GMJF; Sheskin 1982).

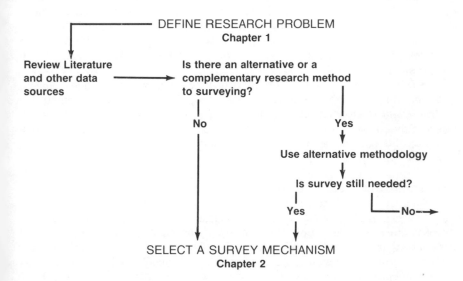

FIGURE 3   DEFINING THE RESEARCH PROBLEM

The Dade County On-Board Transit Survey, a survey handed to bus riders in Dade County, containing a section to be completed on the bus and a section to be mailed in later. The major purpose of this survey was to collect information about the demographics of bus riders and the characteristics of bus trips to assist in realigning Dade County's bus system to act in a 'feeder' capacity to the new METRORAIL system (Kaiser Transit Group 1982; Sheskin *et al.* 1981).

The Southeast Michigan Regional Travel Survey, a home interview survey including a one-hour interview and travel diaries that were left with the household and collected one week later. The major purposes of the survey were to gauge the reaction of the public to various governmental actions that might accompany an energy crisis and to collect information about intraurban transportation to facilitate the urban transportation planning process (Stopher and Sheskin 1982a).

The UM Travel and Parking Survey, a self-administered survey mailed to all faculty and staff of the University of Miami and administered to a random sample of students in a classroom setting. The major purpose of the survey was to obtain information which would indicate the number of University personnel likely to use the University of Miami METRORAIL station. (The study was done about one year before the May 1984 opening of METRORAIL.) The information was used by the University Board of Trustees as a justification for not proceeding with plans for a costly entrance to the University from the station. Additional information from the study was used in a decision to construct a parking lot opposite the station (Sheskin and Warburton 1983).

The Survey of Academic Computing at the University of Miami, a self-administered inventory form distributed to departmental representatives who were subsequently interviewed in a semi-structured format. The major purpose of the study was to elicit information that would be useful in devising a long-range plan for academic computing (Sapp and Sheskin 1983).

The *Miami Herald* Post-Riot Survey, a random digit dialing telephone survey of those Dade County residents with telephones and a home-interview survey of those without telephones. The major purpose of the survey was to assess differences in attitudes toward race relations between blacks living in the inner-city Liberty City riot area, blacks living in a middle-class black suburb, and the general countywide population (Boswell *et al.* 1982).

The Dade County Elderly Mobility Study, a self-administered questionnaire concerning the travel habits of Dade County elderly (Sheskin and Friedman 1981).

The *Miami Review* Readership Survey, a survey of the readers of the *Miami Review* (a metropolitan area business newspaper) designed to examine the level of readership of various items in the paper (Sheskin 1984).

These example surveys are used in this volume because they represent real-world applications of survey research techniques and illustrate a wide range of survey mechanisms, sampling strategies, and logistical concerns. For each survey, the reasons for the selected survey mechanism are discussed at the end of Chapter 2 and the sampling strategy, at the end of Chapter 3. Chapter 4 presents some example questions from four of the surveys. Chapters 5 and 6 incorporate these surveys as examples in discussing the logistics of survey research and the execution of surveys. Brief examples are taken from these surveys to illustrate points throughout the text.

2

Selection of a Survey Mechanism

One of the first decisions in designing any survey is to select the mechanism by which the data are to be collected. Survey mechanisms include:

(1) **Personal interview surveys.** Face-to-face interviews are conducted, usually in the home or work place.
(2) **Mail surveys.** The most common mail survey uses a mail-out/mail-back procedure, but drop-off/mail-back and mail-out/pick-up surveys have also seen use.
(3) **Telephone surveys.** These involve either telephone directory methods or random digit dialing.
(4) **Intercept surveys.** Respondents are approached while in the process of participating in some activity and either are interviewed or are handed a self-administered form.
(5) **Dual survey mechanisms.** These combine any of the above.

Selection from among these alternatives depends on a number of factors, including the purpose of the survey, the definition of the sampling frame (Chapter 3), the required sample size, the types of questions to be asked, the likelihood of obtaining accurate answers, the length of the survey, the spatial distribution of the sample, the educational level of respondents, ethical considerations, labor availability, and time and budget constraints. Each mechanism has advantages and disadvantages in specific situations.

For many years, the most 'acceptable' method was the personal interview survey. One of the classic works in survey research (Backstrom and Hursh 1963) does nothing more than mention that surveys can be completed by telephone or mail methods. In recent years, telephone and mail procedures have gained in popularity, due largely to the influential work of Dillman (1978). The advantages and disadvantages of each of the five mechanisms are now addressed.

The Personal Interview Survey

The personal interview survey involves a face-to-face interview with a respondent, usually in the respondent's home or work place. It has the dual advantage of usually yielding the highest response rate of any survey mechanism and, given effective interviewers, of permitting the use of a rather lengthy survey instrument. The Southeast

Michigan Regional Travel Survey home interview lasted about one hour and achieved an 85% response rate; 2,706 interviews were completed; and only a handful of respondents terminated the interview before completion. Seley (1981) reported an 84% response rate in a survey of businesses. It is much easier for potential respondents to misunderstand the nature/importance of the research and to hang up a phone or discard a letter than to ask an interviewer to leave their home. Personal interviews provide a greater opportunity for the interviewer to explain the importance of the research. In any case, only ethical procedures (see Chapters 4-5) should be used to elicit cooperation.

Ten significant advantages accrue to this methodology. First, unlike a mail survey or a self-administered form, respondents who are visually impaired, illiterate, or not fluent in English may be included; this can be an extremely important consideration in a less developed region (Dixon and Leach 1984). Second, interviewers may explain the meanings of words that might otherwise be misunderstood; if a respondent provides an answer that contradicts earlier responses, the question can be re-read to ascertain whether the respondent perceived it correctly.

A third advantage is that respondents cannot "see ahead" to other questions on the interview form. As an example, suppose the following open-ended question is asked on a mail survey form: "What do you think are the three most important problems facing the residents of this county?" If other questions refer to matters such as taxes and energy policies, these ideas may be suggested to respondents who look ahead. Similarly, on a self-administered form, respondents can look back to see that they had, for example, "slightly agreed" with a question about the comfort of the bus before deciding to "strongly agree" with a later question about the comfort of the automobile.

Fourth, respondents cannot see sensitive questions (such as those involving income) at the end of the questionnaire before answering the remainder of the survey. Some respondents use the fact that such sensitive information is being sought as a reason for not answering the questionnaire at all, even if one explains to them that they are free not to answer any given question. A home interview situation provides an opportunity to explain the reason certain sensitive data are important. In the Southeast Michigan Regional Travel Survey, those refusing to reveal income information were presented with a letter from the 'Survey Director.' This explained the importance of the income question, and asked them to check the appropriate income category on the letter, which they would then seal. This was to be opened by the Survey Director back at the office, where only an identification number linked the envelope with the completed questionnaire, thus protecting confidentiality. This methodology resulted in over 200 responses that otherwise would not have been obtained. Similar methodologies have been used by political scientists in exit polls in which 'secret' ballots are given to respondents which are completed and placed in a ballot box (Backstrom and Hursh-César 1981:23). It is important to remember, however, that respondents should not be forced to answer questions that they do not wish to answer. At no time should a respondent be made to feel unreasonably uncomfortable for the purpose of the research project. Also, if one has promised the respondent that only the Survey Director would open the envelope, this directive must be followed.

Fifth, in the personal interview survey one can use the answer from one question as part of the next. For example, suppose question 1 read: "Which means of transportation did you use the last time you travelled to work?" and the respondent replied, "bus." The next question can then be read as: "How often in a week do you go to work by bus?" rather than the more cumbersome wording that would have to be used on a self-administered form: "How often during a week do you go to work by the same travel means you gave as your answer to question 1?"

Sixth, the amount of missing information and "don't knows" is often much less than for a self-administered form, where respondents often skip questions which they believe are irrelevant to them.

Seventh, one has control over respondent selection. Often, one wishes to interview a randomly-selected household member or the head or spouse of the household head. When a survey is mailed, it may be completed by the wrong household member or even given to a neighbor.

Eighth, open-ended questions (questions for which respondents are not provided with a set of answers from which to choose) are much more likely to elicit a reponse via an interview. In a mail survey, for example, respondents are much less likely to take the time to answer essay questions, or may feel limited by the space provided on the form.

A ninth benefit may be of greater importance to geographers than other social scientists. Only in a personal interview survey can one successfully ask questions related to map interpretation, request drawing of maps, or show photographs (of landscapes, for example) and ask for reactions. Much research in environmental cognition relies on such procedures (Whyte 1977; *Environment and Behavior*). Thus, the spatial and visual nature of some geographic research may dictate a personal interview. The ability to use visual aids extends beyond maps, however. "Response cards" may be handed to respondents showing the choices for some questons. Only in this manner can respondents easily select from more than four choices (such as *strongly disagree, disagree, slightly disagree, neutral, slightly agree, agree,* and *strongly agree*). Halperin and others (1983) effectively used self-administered questions in addition to questions asked by an interviewer in an interview survey of retailers.

A final advantage of the personal interview survey is that, in contrast to mail and telephone surveys, it allows the interviewer to note some information (via observation) about both nonrespondents and respondents, including race, sex, possibly marital status, approximate age, family size, housing type, address, *etc.* This permits some judgment of the effect of nonresponse. It is important to treat information about nonrespondents with the same degree of confidentiality afforded the data collected by the survey instrument.

For all of the positive aspects of interview surveys, eight disadvantages must be considered. First, the interviewer-interviewee situation is contrived. In many ways, for the various reasons described below, surveys create data, rather than collect it. Interviewer appearance, behavior, socioeconomic characteristics, and question-reading style will affect respondents.

For geographers, such interviewer-induced bias may be even more critical when geographic cluster samples (Chapter 3) are used, and, for obvious economies, the same interviewer visits all households in a region. Suppose that a research design calls for comparing attitudes in Regions A and B. Interviewer A is sent to Region A; Interviewer B, to Region B. If regional differences are discerned, it is possible that these differences do not actually exist, but rather are related to differential biases introduced by the two interviewers (Sudman *et al.* 1978).

Another bias of importance to geographers is that response rates are almost always lower in central cities and suburbs than in rural areas. Fowler (1984:50) suggests three reasons for this. First, central cities are home to many difficult-to-locate single persons. Second, interviewers have problems gaining access to respondents in high-rise buildings. Third, more areas in central cities exist where interviewers are uncomfortable at night. Thus, there is likely to be spatial bias to the nonresponse in a personal interview survey.

Other biases are introduced when respondents try to impress or 'help' the interviewer by providing answers they think the interviewer wants to hear, or by providing an answer when a 'don't know' is more appropriate. The former factor was instrumental in the decision not to use a home interview in the GMJF Demographic Study: interviewees might overstate their level of religiosity in an attempt to please or impress the interviewer.

Additional bias may be introduced in a home interview survey when other household members observe the interview. For example, suppose a survey has been designed to elicit trip information from a randomly-selected household member. If a husband is questioned in front of his wife, he may tell the interviewer about his trip to work and to the grocery store, but not to the local bar!

A second disadvantage of personal interview surveys is that they are highly labor intensive and, therefore, quite expensive. Because of the need for repeated call-backs at households where no one was at home or where the respondent was too busy when originally called upon, interviewers are unlikely to complete more than 2-3 interviews in a three-hour evening shift. In recent years, labor costs have skyrocketed because the large, inexpensive female labor pool available in the post-World War II era has evaporated. Contacting respondents has become more difficult due to changing life styles, including increases in the percentage of households containing two employed spouses or only one adult. Another contributing factor is the increase in the percentage of meals consumed in restaurants.

A third disadvantage is that some respondents are unreachable because guards limit entry to luxury condominiums and other high-rise developments. In Dade County, a guard is stationed at the entrance to the island municipality of Golden Beach, and only residents (or their guests) are allowed in after dark.

Fourth, from an urban geographer's viewpoint, a significant problem exists with the spatial coverage afforded by a home interview survey. Almost invariably, the sample is drawn such that blocks are sampled first; 3-5 housing units then are selected on each block. Thus, as many as five housing units may have almost exactly the same location with respect to major activity centers in a city. In addition, this procedure reduces data variation because households located on the same block are more likely to be similar than those that would be drawn for a random sample of all houses in an area.

Fifth, home interview surveys take many more months to complete than telephone or mail surveys. Sixth, personal interviews may be impossible with a population that is widely dispersed, such as dentists or university alumni.

Seventh, a significant problem can arise if interviewers are not totally trustworthy. In the Southeast Michigan Regional Travel Survey, a few interviewers were caught falsifying entire survey forms. This cannot happen in a carefully controlled telephone survey, for example.

A final problem with the home interview survey relates to ethical considerations. Of all the various survey mechanisms, the home interview creates the greatest number of problems: a strange interviewer arrives at the door (often without advance warning) and requests permission to enter and ask questions (sometimes *sensitive* questions) for perhaps one hour or more. The telephone, mail, and intercept surveys are significantly less intrusive. Related to this problem is that the interviewer in a home interview survey is almost always in possession of the name and/or address of the interviewee and potential problems arise with respect to protecting confidentiality. In most other types of surveys, the interviewer only has access to an identification number for each respondent.

Another ethical problem arises because this survey mechanism necessitates that some interviewers visit 'dangerous' areas and enter strangers' homes. No interviewer should be required to promise to conduct interviews in such areas as a condition of employment. In the Southeast Michigan Regional Travel Survey, one interviewer was faced with a potential interviewee holding a gun; another 'gentleman' answered the door naked and would agree to be interviewed by the female interviewer only in this state. After these incidents, none of the interviewers would return to this neighborhood. The important point is that, if a home interview survey is to be conducted, rather than a telephone, mail, or intercept survey, solid arguments should be available to support this decision, given the above ethical considerations.

The Mail Survey

This survey mechanism involves the use of a self-administered form which usually is mailed out to, and then mailed back by the respondent. It is the least expensive survey mechanism and has the simplest logistics, as there is no crucial "survey day" around which all plans need be coordinated. It has a relatively small labor requirement and can be accomplished at a comparatively low cost.

Five additional advantages may be cited for mail surveys. First, the absence of an interviewer eliminates the interviewer-induced biases that can plague personal interviews. Second, because interviewer time and mileage are not of concern, a cluster sample is not necessary, thereby improving the spatial coverage of the sample.

Third, because respondents can see the questions and answers, more complicated questions can be incorporated in a mail than in a telephone survey.

Fourth, a mail survey provides an opportunity to consult with other household members on fact questions. For example, suppose a recreational geographer was studying whether or not Chicagoans use Lake Michigan. If asked via personal or telephone interview, respondents are unlikely to admit that they do not know where Lake Michigan is, and are unlikely to have an opportunity either to check with other household members or to consult a map to see whether or not the beach or marina they visit is on the lake.

Fifth, a mail questionnaire gives respondents time to think about questions. Often in an interview situation, interviewees are reluctant to 'hold up' the interview to give serious consideration to questions. Finally, mail questionnaires do permit maps and other visual cues to be included.

Eleven disadvantages to the mail survey may be noted. The major disadvantage is that the response rate is likely to be lower than for interview surveys. Mail surveys traditionally have yielded the lowest response rates of any survey mechanism. For many years, both mail and telephone surveys were considered the "stepchildren of survey research" (Dillman 1978:1). Recently, the mail survey has achieved a new level of respectability, not in any small part due to the work of Dillman (1978) and his "total design method" (TDM). Dillman reports an average response rate of 74% for 48 TDM mail surveys; no survey achieved less than a 50% response rate. In the GMJF Demographic Study, using many of Dillman's TDM procedures, over 79% of the mail surveys were returned. However, a number of studies by geographers, using mail surveys in examining the locations and the decision-making process of industrial firms, have reported low response rates: Scott (1983, 1984) achieved response rates of 11% and 9% in an examination of the printed circuits and the women's dress industries in Los Angeles; Wheeler (1981) achieved a 32% response rate in his study of manufacturing in Georgia. Talarchek (1982), in a study of household residential search behavior, achieved a 46% response rate.

The low response rates sometimes achieved on mail surveys are of particular concern. To a much greater extent than any other survey mechanism, those who have an interest in the subject matter, or those who feel strongly about an issue, positively or negatively, are more likely to respond. This means that mail surveys almost always will be biased significantly in ways that are related directly to the purpose of the research (Donald 1960).

Comparable response rates between home interview and mail surveys generally have not been reported. Thus, much of the supposed advantage of the personal interview survey in terms of response rate may be exaggerated. Whereas response rates for home interview surveys often are cited in excess of 90%, such rates are generally misleading, as they are calculated on a different basis than mail survey response rates. For the latter, response rates are calculated as the proportion of mailed surveys returned as usable responses. Frequently, an unknown proportion of mailed surveys are not delivered, or are delivered to addresses that are temporarily or permanently vacant. Interview survey response rates usually are based on the total number of completed interviews, plus terminations and refusals; not included are those situations in which an interviewer finds no one at home, a site under construction, "no such address," or fails at a request for a call-back. These are normally ignored in calculating a response rate and respondents from a back-up sample are substituted.

The reporting of comparable response rates between home interview and mail surveys would show interview surveys achieve a much lower response rate than is usually claimed. As an example, for the Southeast Michigan Regional Travel Survey, the calculated response rate was 85%. If survey attempts which resulted in "no one home" and the like are considered, the response rate drops to 65% (Stopher 1982). This idea is consistent with Dillman's argument that "in face-to-face and telephone interviews a refusal is not considered as such until a contact is made. In mail studies, the opposite is assumed, that is, a nonresponse is a refusal until proven otherwise" (1978:50).

Although it is still likely that a higher response rate will be achieved for a home interview than for a mail or telephone survey, if a mail survey is executed well, incorporating appropriate features to increase the response rate (Chapter 5), and if response rates are calculated in a comparable manner, the disparity will not be as great as has traditionally been assumed.

Response rates from home interview surveys have been declining in recent years as interviewers find increases in the "no answer" category (for reasons discussed above) and a reluctance to permit entrance due to a fear of crime. In the GMJF Demographic Study, the home interview procedure was rejected for the latter reason.

Several other reasons for the decline in response rates for home interviews are discussed by Backstrom and Hursh-César (1981:47). First, many people do not trust polls because some highly-publicized polls have been incorrect. Second, some people complain about oversaturation of surveys, particularly in university towns, although most surveys are done by commercial firms for commercial reasons. In addition, the public is beginning to have an adverse reaction to "snooping" by various agencies and the keeping of records by computer. Third, many sales efforts misrepresent themselves as surveys. Finally, surveys are often undertaken for political candidates and only positive results are released. Likewise, some "man-on-the-street" interviews are represented as surveys. These factors have undermined public confidence in surveys, and reduced the response rates for all surveys, regardless of survey mechanism.

A second disadvantage of a mail questionnaire is that respondents who are visually impaired or illiterate are unlikely to respond. Third, respondents can see ahead to other questions. Fourth, question wording can often be complicated because the answers to previous questions cannot be anticipated. Fifth, many returned mail surveys contain a significant amount of missing information. (This may be correctable using a follow-up telephone call.) Sixth, mail surveys often are put aside with good intention but are then forgotten or are discarded as junk mail.

Seventh, respondents can only be selected if they appear on some address list. If the universe to which one wishes to infer is, for example, all residents of a county, it is usually impossible to obtain a complete address list. Eighth, because respondents can see the length of the questionnaire prior to deciding whether to participate, great care must be taken to keep the questionnaire as short as possible (although Dillman [1978:55] provided some evidence that length is not a problem as long as the questionnaire does not exceed 12 pages).

Ninth, some respondents have difficulty expressing themselves in writing and open-ended questions are a problem. They also are much less likely to 'stick with' a tedious question. In the GMJF Demographic Study, the greatest occurrence of item non-response on the mail questionnaire involved a question in which respondents were asked which of four responses represented their "degree of use" of 36 different social services.

A tenth problem may exist if respondents need to be located within an office or agency. For example, for the *Miami Review* Readership Survey, it was necessary to locate a reader of this business newspaper within each office receiving the paper. If a mail questionnaire had been addressed to "any reader of the *Miami Review*," the response rate probably would have been dismal. Although the envelope could have been addressed to the "president" of the company, with instructions to ask a reader of the newspaper to complete the form, this would have placed the questionnaire in double jeopardy: either the president or the person to whom it was passed could have discarded it.

Finally, suppose a questionnaire contains different sets of questions (each set occupying one page) for persons who take the bus, the train, and the car to work. Three pages are needed even though each respondent need answer only one. This makes the questionnaire appear considerably longer and discourages response. Moreover, many respondents experience difficulty following the 'skip patterns' necessary to direct them to the proper set of follow-up questions, such as: "If no, go to Question X" and, "If yes, go to Question Y."

The Telephone Survey

The telephone survey has characteristics in common with both the personal interview and the mail survey and combines, for many applications, the best of both. Response rates on telephone surveys are usually quite high. Dillman (1978), using TDM telephone surveys, reported an average response rate of 91% for 31 surveys. Response rates may vary greatly, however. In the GMJF Demographic Study, a response rate of 75% was achieved, with most non-response occurring in areas containing the elderly poor. In the *Miami Review* Readership Survey, over 97% of those respondents reached cooperated with the survey.

Interview length can be longer than for a mail questionnaire, but usually should be shorter than for a home interview. Backstrom and Hursh-César (1981:19) suggested that a twenty minute call is a long interview and thirty minutes is the practical maximum. Dillman (1978:55) reported that, for an interview averaging about 20 minutes, only 10 out of 1018 respondents, drawn from the general public, terminated the interview in the middle. Concern exists among researchers that the quality of information obtained toward the end of a lengthy telephone interview is suspect. However, Groves and Kahn (1979) showed that, for the most part, the quality of data obtained via telephone and personal interviews are comparable.

A number of advantages of telephone surveys may be cited, some of which have been discussed above. First, one has control over which household member is interviewed, and there is less chance of contamination of the answers by other household members. Second, open-ended questions can be asked effectively, although some problem exists in recording the answers quickly enough so that the respondent is not listening to a "dead" telephone for too long.

Third, interviewers may be trained to follow skip patterns on the questionnaire. Skip patterns cause problems on any self-administered questionnaire because respondents, regardless of educational level, often fail to read instructions.

Fourth, nonresponse to questions is likely to be even lower than for a personal interview and much lower than for a mail survey, except for sensitive questions. Whereas a respondent may be more likely to provide income information over the telephone due to the increased anonymity, there is not quite as much anonymity as in a mail survey. In the pilot study for the GMJF Demographic Study, both a telephone survey and a telephone/mail dual survey mechanism were tested. Ten of 23 respondents refused to reveal income data over the telephone, whereas only 1 of 25 refused such on the mail survey. On the whole, however, sensitive information is willingly revealed in telephone surveys. Backstrom and Hursh-César (1981:141) showed that, for four national telephone surveys, only 4% refused to reveal income; 3.5%, political party affiliation; 0.4%, religion; and 3.3%, age.

Fifth, the amount of time needed to complete a telephone interview survey is relatively short. A telephone survey of 400 respondents can often be completed easily within one week with ten interviewers. Completing 400 personal home interviews, using 10 interviewers each completing about 15 interviews per week, would probably require more than three weeks. Whereas most mail questionnaires are usually returned within about 10 days, the necessity to mail follow-up questionnaires to non-respondents and then perhaps to call them, may result in as many as eight weeks for complete data collection.

Sixth, in common with the personal interview survey, visually-impaired or illiterate respondents can participate, and respondents cannot "see ahead" to other questions.

Seventh, while the effect of the personal appearance of the interviewer is obviously not an issue in a telephone survey, question-reading manners can still influence answers. Such factors as answering to please or impress the interviewer are still present, although probably less likely to have a significant effect than is true for a home interview survey.

Eighth, a logistical point in favor of telephone interviews is that it is often much easier to find interviewers. Only persons with available transportation can generally administer at-home interviews. Also, interviewers hired during the course of a telephone survey can be trained by listening to interviews conducted by those already trained and, because of the productivity rate in a telephone survey (2-4 per hour depending survey length), a smaller staff is needed.

Ninth, the survey director can be present at every interview and can immediately remedy ineffective procedures. Although many researchers use a procedure in which the supervisor monitors calls from another telephone, this procedure is viewed by many as unethical, even if the respondent is informed of the situation. It is common in personal interviews for interviewers to take liberty in reading questions and interpreting answers. It is also possible in personal interview surveys for interviewers to fabricate interviews, especially if payment is made for each completed interview. These problems can be eliminated in a telephone survey.

Tenth, an important advantage for geographers is cited by Fowler (1984:50) who reported that "there is some evidence that telephone procedures may reduce the differential response rate between central cities and rural areas because it is possible to give thorough coverage to urban households, to make contact with people in high-security buildings, and to make a very large number [sic] of efforts to find single people at home." Thus, the spatial bias in nonresponse between central city and rural areas that exists in many personal interview surveys is usually less pronounced in a telephone survey.

A final advantage of the telephone survey is that, because of the lower labor costs, it is significantly less expensive than the home interview.

Three disadvantages of the telephone survey may be noted. First, one has to have a telephone to be included in the survey. About 96% of U.S. households have a telephone (U.S. Department of Commerce 1984: Table 949), although this percentage varies significantly from place to place. Roseman and Williams (1980) found that 83% of households in the urban portion of their study area had telephones, but less than 70% did so in the rural areas. Within an urban area, telephones are likely to be ubiquitous in middle and upper class suburbs, but less common in poor, inner-city neighborhoods. For the *Miami Herald* Post-Riot Survey, it was believed that this fact would bias the results and a home interview procedure was used to supplement the sample. For the GMJF Demographic Study, although it was recognized that many of the elderly living in nursing homes and retirement "hotels" would be excluded, a decision was made to accept this bias, given the purpose of the survey. The percentage of households with telephones also varies significantly by state: only about 83% of households in Mississippi and Nevada had telephones in 1981, compared to practically 100% in Connecticut, New Jersey, and Vermont.

It is also obvious that telephone surveys would have very limited utility in most developing countries where telephone ownership is restricted severely by both geographic location and social class. In Israel, a telephone survey would omit persons who recently have changed dwellings because it takes more than two years to have a telephone installed. People who live on kibbutzim have few telephones. Dixon and Leach (1978b:17) suggested that "the North American telephone interview would probably meet with great suspicion in Britain," and, thus, a much lower response rate might be expected.

A second disadvantage of the telephone survey is that it is often difficult to find some respondents at home. This is a problem for travel surveys in particular, as those persons who travel more are less likely to be at home. This bias can be somewhat ameliorated by calling each respondent at least five times, at different hours of the day and during at least two different weeks. Dillman (1978:47) provided evidence which suggests that a mail survey follow-up can be a successful way of eliciting responses from households that repeatedly do not answer the telephone. A related difficulty is that secretaries may not permit the interviewer to speak with the selected respondent.

A final problem with telephone surveys is that respondents have trouble following a question with more than four choices and it is nearly impossible to ask respondents to select, for example, the three most important items from a list of ten.

The Intercept Survey

An intercept survey involves either the distribution of a self-administered form or the personal interviewing of users of a given facility. It is a particularly useful mechanism for surveying travel behavior, shopping behavior at particular retail outlets or malls, use of recreational and other facilities (Tourism and Recreation Research Unit 1983), and voting at the polls (Backstrom and Hursh-César 1981). Response rates for such surveys vary significantly. If a self-administered form is used, the response rate is likely to be lower than if an interview procedure is employed. Also, to the extent that people 'sort themselves out' in the use of transport and other facilities, the population surveyed varies greatly from one intercept survey to another. For example, for the Dade County On-Board Transit Survey, a self-administered intercept survey, only about 23% of eligible bus riders completed the form. This low rate was probably due to the presence of many elderly persons, tourists and part-year residents, Haitians (no Creole survey form was available), Hispanic immigrants unfamiliar with surveys, and people whose dubious legal status in the United States make them reluctant to cooperate. In a similar survey completed for the Florida Department of Transportation, response rates of 60-80% were achieved in Bradenton, Pensacola, Daytona Beach, and Orlando, while Dade County recorded only a 34% response rate (Schimpeler.Corradino Associates 1983). In the Ann Arbor-Ypsilanti Area Transportation Study (Schimpeler.Corradino Associates 1980), over 80% of on-board transit surveys were returned, probably due to the unique environment of this university area.

The Tourism and Recreation Research Unit (1983) in Britain reported that it achieves response rates in excess of 90% for most on-site personal interview surveys. For self-administered forms, a two-thirds response rate is considered good, with many surveys attaining very low response rates. A survey was designed by Cambridge Systematics Inc. (1984) to examine the impact of the Detroit downtown people mover on the CBD travel behavior of persons staying in Detroit hotels by handing questionnaires to respondents as they checked in. The response rate was well below 10% and the results could not be used. Obviously, travellers had little motivation to participate. For the UM Travel and Parking Survey, students were intercepted in their classes and self-administered forms were used. Because of this special controlled situation, a 100% response rate was achieved.

Thus, response rates to such surveys depend on the type of facility being surveyed, the type of population using the facility, the methodology of the survey, and the size of the city.

The permissible length of the intercept questionnaire depends, to a large extent, upon the circumstances under which it will be completed. For a self-administered form that is to be taken home and mailed in, brevity should be the rule, as in a mail survey. For a self-administered form that is to be completed during some activity, the form should be kept especially brief. For example, the Dade County On-Board Transit Survey encouraged respondents to complete the form during their bus trip. Thus, the length could not exceed two pages. In the UM Travel and Parking Survey, the need to avoid significant interference with planned class activities dictated that this survey also not exceed two pages.

For a personal interview intercept survey, length depends on the location of the interview; brevity is prompted by many such situations (Figure 4). The Tourism and Recreation Research Unit (1983) recommends either a two-minute short questionnaire or a 5-10 minute longer questionnaire. In the survey of mode-of-access to METRORAIL in Dade County, it was decided that respondents would have time to answer only one question as they hurried to catch the train.

Most of the advantages and disadvantages of personal interview and mail surveys apply to interview intercept and self-administered intercept surveys and will not be repeated. Four additional advantages of the intercept survey may be cited. First, it may be the only cost-effective procedure for reaching a particular population. For example, in Dade County only 4% of the population use the bus system. Thus, for a bus ridership survey, 96 telephone calls of every 100 would *not* yield an eligible respondent. Second, because respondents are being asked about a particular behavior while engaged in it, they are less likely to forget information. In transportation studies, respondents often forget trip details when asked about them later in the day. Third, interviewers can observe certain information about respondents, even in conjunction with a self-administered form. Depending upon traffic flow at the survey location, interviewers can record the sex, race, approximate age, automobile license plate number, direction of travel, and behavioral characteristics of each potential respondent (in a recreational survey, for example, did the person appear to be a fisherman, swimmer, or boater?). This permits some checking on the characteristics of nonrespondents and allows certain information to be added to each survey record without the need for asking the question on the survey form. Finally, a significant advantage to geographers is that respondents can be asked to evaluate an environment in which they are currently located.

One important drawback exists to interview intercept surveys when potential respondents pass the interview point more quickly than the time needed to complete each interview. The best procedure to use to select a respondent is the "next-to-pass" rule. That is, when a survey is completed, then the next person to pass the interview point is asked to be a respondent. If a group of persons is next to pass, then the one whose birthday is next is interviewed. A drawback is that the interviewer is actually selecting the sample and it is likely that interviewers may be subconsciously tempted to approach persons who are ethnically similar and who are sexually different, particularly when traffic flow is great.

"I'm from the M.T.A. Do you feel that the B.M.T. is more comfy than the I.R.T.? How does either of them compare, in your opinion, with a bus? Would you rather be riding a taxi, or even driving your own car?"

FIGURE 4   A TRAVEL SURVEY. Drawing by Mandelsman; ©1982 *The New Yorker* Magazine, Inc.

The Dual Survey Mechanism

The dual survey (also called a mixed-mode survey) combines two or more of the four procedures discussed above, and allows one to profit from the advantages of each of the individual survey mechanisms, with the advantages of one often negating the disadvantages of another. Four advantages may be cited.

First, respondents are more likely to respond to either a brief intercept survey or a 5-10 minute telephone interview than to a significantly longer survey of any type. For many, a sense of involvement is created which leads them to complete the subsequent longer survey — the standard marketing technique that compliance with a small request leads to compliance with a subsequent longer request (Bem 1972). Second, the initial survey can be used to obtain an address to which follow-up reminders may be sent, increasing the response rate for the second part of the survey.

Third, because the response rate should be much higher on the shorter part of the survey, this information can be used to ascertain any response biases for the longer mail questionnaire. For example, if 60% of those returning an intercept survey or answering the telephone survey are elderly, but only 30% of those returning the mail survey are elderly, then weighting factors can be added to the mail surveys to account for the under-representation of the elderly.

Fourth, if one of the two surveys involves an interview and the other is self-administered, questions can be placed on the interview form which are best asked via interview. Among these are questions that are open-ended, have difficult skip patterns, require the use of pronouns, must be answered before other questions are seen, and necessitate long follow-ups that apply only to a portion of the sampled population. Alternatively, questions best asked on a self-administered form can be placed there. These include questions listing many answers, involving ranking of alternatives, requiring considerable thought, using map and other visual aids, addressing sensitive issues, and necessitating that other household members be consulted. Also, important questions can be placed on the survey that will yield the greater response rate.

As an example of a dual survey mechanism, Figure 5 shows the rather complicated procedure used in the GMJF Demographic Study. Respondents were first sent a post card (1) indicating that they would receive a phone call. When called (2), if a respondent seemed to be refusing (3) a *very* quick "Survey of Last Resort" (which collected minimal information) was attempted (4). If this failed, a "Refusal Report" (5) (sex, approximate age, reason for refusal) was completed. If respondents refused to answer questions on the telephone, they were offered the option of having all the questions sent to them (8). If respondents completed the telephone survey (2), they were informed that they would be sent a mail survey (6). If they declined to be sent a mail survey, they were offered the option of answering the mail questionnaire over the telephone (7).

After three days, all respondents were sent a postcard either thanking them for returning the survey or reminding them to do so (9). If no response was achieved in three weeks, a second copy of the mail survey was sent (10).

After another five weeks, all nonrespondents were called (11) and either refused (12), were reminded to mail the survey (13), were sent an additional copy (14), or were asked the questions over the telephone (15). Those who were reminded or sent an additional copy were called again in another two weeks if necessary (16). Finally, respondents who had missed a significant number of questions on the mail survey were called (17) to complete the missing information.

Although some respondents answered certain questions over the telephone that others answered through the mail, this created less bias than if the former households had to be classified as nonrespondents. Because the survey was designed logistically to 'work' as either a telephone or mail survey, the preferences of different respondents could be met, and the response rate maximized. A 75% response rate was achieved on the telephone portion and a 79% rate for the mail portion.

The Eight Example Surveys

The reasons for selecting the survey mechanisms for each of the example surveys follows. Note that the basic purpose of each survey is presented at the end of Chapter 1 and the sampling strategy for each at the end of Chapter 3. Chapter 4 presents some sample questions from four of the surveys. Chapters 5 and 6 incorporate these surveys

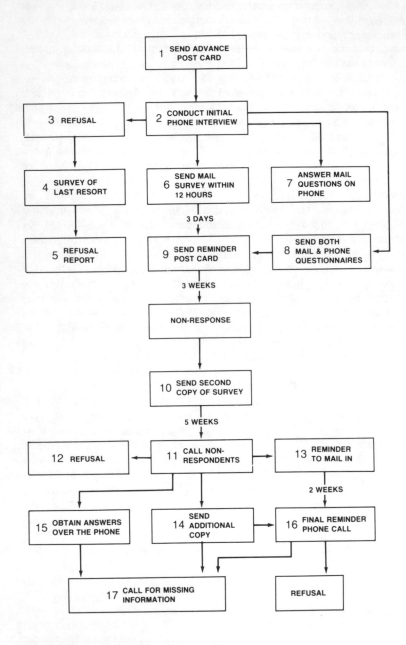

FIGURE 5   TELEPHONE/MAIL DUAL SURVEY MECHANISM
(Sheskin 1982:11)

as examples of survey logistics and execution.

For **The GMJF Demographic Study,** the home interview would have been cost prohibitive because of the necessary sample size, and the response rate might have been very low among those elderly afraid to permit strangers into their homes. Interviewer bias would have led to unreliable answers to questions about religiosity. The telephone and mail-out/mail-back procedure seemed most appropriate.

For **The Dade County On-Board Transit Survey,** only an intercept procedure could effectively locate bus riders, who comprise only about 4% of the population. However, the information required was too extensive to be completed while riding the bus. Thus, a dual survey mechanism was used in which there was a personal request to complete and hand back a short questionnaire form on the bus, in combination with a longer take-home/mail-back survey.

For **The Southeast Michigan Regional Travel Survey,** a one-hour interview needed to be followed by an explanation of a travel diary distributed to all household members. This could most effectively be accomplished in person.

For **The UM Travel and Parking Survey,** the most efficient manner for delivering self-administered forms to faculty and administrators was with a mail-out/mail-back mechanism. For staff, supervisors handed questionnaires to people at their desks or in groups. Students were intercepted in randomly-selected classes.

For the **Survey of Academic Computing at the University of Miami,** the open-ended nature of all the questions dictated the need for a personal interview. Because all interviews took place in faculty offices on one of three campuses, most of the logistical problems inherent in personal interview surveys were avoided.

For **The *Miami Herald* Post-Riot Survey,** results were desired within three weeks and a very limited budget was available, necessitating the use of a telephone survey with some home interviews of persons without telephones.

For **The Dade County Elderly Mobility Study,** a self-administered questionnaire was given to groups of elderly participants at activity centers. The budget for this pilot project did not allow for the usage of any other mechanism.

For **The *Miami Review* Readership Survey,** a five-minute telephone interview was administered. A mail survey addressed to "any reader" of the paper at a business address was unlikely to be returned. Because readers of the paper were not likely to be geographically clustered, a personal interview survey would have been overly costly. Also, the subscribers to this publication were much more likely to agree to answer five minutes of questions on the telephone than to agree to be interviewed in person during business hours.

Which Method is Best?

The answer to this question is that "It depends." Instances exist in which one method is clearly the only "workable" procedure. If one were to document the journals read by professional geographers in 50 countries, the only reasonable method would probably be a mail survey. If only two weeks were available in which to collect data, a telephone survey would be the likely choice. If a very long questionnaire was needed, containing many complicated procedures, an interview survey would be indicated. If one needed to interview persons utilizing a recreational facility, only an intercept survey would usually be feasible. Due to the greater likelihood of a low response rate, a mail survey is not recommended unless it is part of a dual survey mechanism, it is aimed at a specific audience (such as a survey of nurses about nursing), or it is the only reasonable procedure for obtaining the needed information.

Sometimes the decision as to which survey mechanism to use is determined by an overriding factor, such as cost or time constraints, while at other times, the decision is not an easy one. A good procedure to follow, when some doubt exists, is to pilot test each possible survey mechanism prior to making a determination. This was done for the GMJF Demographic Study, resulting in the use of the telephone/mail dual survey mechanism rather than a telephone-only approach. The choice of a survey mechanism may be influenced also by a number of issues related to sampling, the subject of Chapter 3.

3

Sampling Issues

Even if a researcher has unlimited time and resources, there is little cause to gain information from *all* members of a population. Sampling is well-established and accepted, both within the scientific community and by the general public. The *goal* of sampling is to select a small number of 'representative' individuals from a population (Kalton 1983; Mendenhall *et al.* 1971; Slonim 1960). Survey respondents usually are selected in some random or pseudo-random manner, but, for a variety of reasons, a truly random sample of respondents is rarely achieved. This chapter discusses the issues of sample size, sampling frames, and sampling designs (Figure 6). The decision-making processes used to resolve each of these issues for the eight example surveys presented in Chapter 1 is then addressed.

This discussion assumes that the reader has a beginning knowledge of introductory inferential statistics and makes no attempt to provide the necessary theoretical background for its conclusions. Readers are referred to Blalock (1979) and Dixon and Leach (1978a).

Sample Size

An important issue in any survey design is the determination of sample size. This issue is related to the choice of a survey mechanism (Chapter 2). A particular survey budget may be able to accommodate 400 personal interviews, but another methodology may be dictated if 1600 respondents are required. Five major factors play a role in the determination of sample size: cost, time, geography, level of accuracy, and subgroup analysis.

An important relationship exists between costs and sample size (Figure 7). Survey costs may be divided into "fixed costs," which do not vary with sample size, and "variable costs," which are related directly to sample size. Fixed costs, represented by Point A on Figure 7, include the amount of time needed for a professional to design the questionnaire, sampling strategy, and data analysis, and to write the report. Variable costs include worker and supervisor time for an interview survey, postage for a mail questionnaire, and printing costs for any type of survey. Variable costs increase relatively quickly at first, but the rate of increase 'tapers off' as economies of scale in printing, interviewer training, and other items take effect. For many surveys, fixed costs are the major portion of the budget. Increasing the sample size in a telephone survey by 100, for example, might only imply small increases in the interviewer payroll and coding and keypunching charges.

Another concern is the amount of time available to complete the survey, although

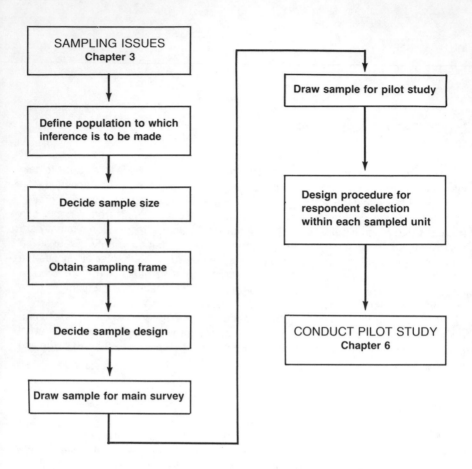

FIGURE 6   SAMPLING ISSUES AND DESIGN

this problem can be ameliorated by hiring additional labor. Nevertheless, if the time allocated to a survey is short, some compromise on sample size may be necessary.

Usually, the more geographically dispersed a population is, the larger the necessary sample size. Populations distributed over different types of environments are likely to be more variable with respect to demographics, attitudes, and beliefs.

A major factor in determining sample size is the selection of a desired accuracy level for the survey results. It is common for surveys to report both percentage responses and numerical averages. The nature of these results also affect sample size. The most commonly-selected accuracy level is the situation in which one is 95% certain that no estimated percentage is off by more than $+/-5\%$ (referred to as '95 and

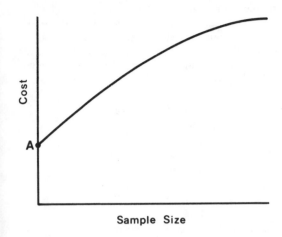

Sample Size

FIGURE 7   RELATIONSHIP BETWEEN SAMPLE SIZE AND COST

5'). 95% is the confidence level; 5% is the confidence interval. The sample size **n** necessary to achieve such a result is computed as follows:

$$n = \left(\frac{Z \sqrt{PQ}}{C} \right)^2 \qquad (1)$$

where: Z = 1.96, for 95% confidence that a result lies within a given confidence interval

Z = 2.58, for 99% confidence that a result lies within a given confidence interval

P = the percentage about which a confidence interval is computed, expressed as a proportion

$Q = 1 - P$

C = the desired size of the confidence interval, expressed as a decimal number

A simple algebraic examination of equation 1 suggests that:

(1) A 99% confidence level requires a larger sample than a 95% level.
(2) As the size of the acceptable confidence interval increases, the size of the sample needed decreases.
(3) If a percentage is split 90%-10% (for example, 90% of respondents answered "yes" and 10%, "no"), a smaller sample size is needed to achieve a given confidence interval at a given confidence level than if, for example, the proportion is divided 50%-50%.

Imagine that a decision has been made that 95 and 5 is an acceptable confidence level and that the major survey purpose is the estimation of the percentage of people using the bus for the work trip (called the "modal split"). A problem arises because estimating the necessary sample size to achieve 95 and 5 is dependent on the value of a percentage that will not be known until after the survey is completed. Three "solutions" to this predicament are:

(1) Use information from the pilot study to estimate the likely percentage that will be found in the main survey.
(2) Use information from other studies concerning the likely modal split value. With the exception of New York City, very few American cities exhibit modal splits higher than 10%. One would be relatively safe in suggesting that **P** be set equal to .1 and **Q** to .9
(3) Assume the worst-case situation. In equation 1, **n** is largest when **P** = .5 and **Q** = .5. Thus, most researchers simply substitute .25 for $\sqrt{PQ}$. This procedure also guarantees that every percentage calculated from the survey will be *at least* 95 and 5.

Most surveys are designed using the third method. Table 1 shows the necessary sample size to achieve various confidence intervals for the 95% and 99% confidence levels. Larger sample sizes are needed as the confidence interval narrows or as the confidence level increases. Most importantly, note that halving the confidence interval (from +/−6% to +/−3%) requires quadrupling the sample size (from 267 to 1067). Also, increasing the confidence level by only 4% (from 95% to 99%) implies a 74% increase in sample size. Table 1 shows that the popular '95 and 5' option requires a sample size of 384.

Although most geographic surveys involve sampling from large populations, some require sampling from small populations. Equation (2) is used to adjust the values in Table 1 for small populations:

$$n^* = \frac{n}{1 + \left(\dfrac{n}{N}\right)} \tag{2}$$

where: **n***   = necessary sample size when sampling from a finite population
          n   = the sample size from equation (1)
          N   = the population size
          n/N = the sampling fraction

Note that the sampling fraction (the percentage of the population to be interviewed) has little effect on sample size. For example, if a sample is drawn from a population of 100,000 (implying a sampling fraction of 384/100,000 = .4%), equation (2) indicates that a sample of "only" 383 (instead of 384) is needed for 95 and 5. If the population size is 5,000 (sampling fraction = 384/5,000 = 7.7%), a sample of 357 is needed.

TABLE 1 NECESSARY SAMPLE SIZES TO ACHIEVE VARIOUS CONFIDENCE INTERVALS[a]

Confidence Interval	Confidence Level 99%	95%
1%	16587	9604
2%	4147	2401
3%	1843	1067
4%	1037	600
5%	663	384
6%	461	267
7%	339	196
8%	259	150
9%	205	119
10%	166	96
15%	74	43
20%	41	24

[a](Assumes $P = Q = .5$ and a very large population)

A very important concept is that once a population consists of 100,000 or more, equation (2) has no effect on sample size. Even for populations below 100,000, the sampling fraction has little influence until the population is relatively small. Thus, whether a geographer is surveying a city of 100,000 or 10,000,000, the 'magic' number is still 384. In practice, most researchers attempt to obtain about 400 completed responses; usually some results must be discarded during data analysis.

In some surveys, the most important information collected will result in the calculation of an average, rather than a percentage. In such a case, the sample size is calculated from:

$$n = \left(\frac{ZS}{C}\right)^2 \qquad (3)$$

where: Z, C are as above
$S =$ the standard deviation of the average

This equation has a similar problem as equation (1): the standard deviation is not known until after the data are collected. Again, information from a pilot survey or from other studies could be used, but there is no procedure equivalent to assuming $P = Q = .5$. This has been a particular problem for transportation surveys, in which the most important variable is the average trip rate per household. Because no knowledge of the standard deviations of trip rates was available during the 1950's and 1960's, transportation surveys used samples of 1 to 5% of the regional population, with up to 30,000 households interviewed. With such knowledge available since the 1980s, the Southeast Michigan Regional Travel Survey used a 0.16% sampling fraction (2,704 households). In a 1965 southeast Michigan travel survey, a 4% sampling fraction was used. Had this sampling fraction been used in the 1980 survey, 66,000 households would have been interviewed and the survey cost would have been $6.3 million higher (Stopher 1982).

A final factor in sample size determination concerns the population subgroups to be analyzed. Assume that a telephone survey has been designed and that sufficient budget and time exist to achieve a 95 and 5 accuracy level. Thus, a sample size of 400 is selected. The next step is to identify the population subgroups for which individual statistics are required. Suppose that the elderly is one such group, and that it is known *a priori* that 12.5% of the population is elderly. In a typical sample, then, approximately 50 respondents will be elderly. Table 2 shows that results for the elderly will have a confidence interval of 13.9%. For some applications, this might be unacceptable. If one desires the level of accuracy for the elderly to be 95 and 5, then a sample of 3200 would be needed to obtain 400 elderly responses (probably an unjustified expense in most cases). Alternatively, the survey could be executed as planned and then a second survey (using a sampling frame of elderly residents) could be undertaken to supplement the sample.

As a second example, many transportation surveys produce data for cross-classification trip generation analysis. Florida has a standard cross-class model for which urban areas produce their own data (Figure 8). To use the model, data are needed on the trip-making behavior of 30 different types of households, defined by dwelling unit type (single family, multi-family), persons per dwelling unit, and autos per dwelling unit. The survey must have a sample size sufficient to include a satisfactory number of households for most of the 30 cells. One may define a 'critical cell' in which it will be most difficult to find respondents. If only a few percent of respondents 'fall' in this cell (or population subgroup), obtaining a sufficient number of households in the critical cell will determine the necessary sample size. In actual practice, the number of households in some cells is so small that these cells may be ignored. When conducting telephone interviews, the first three questions should establish dwelling unit type, number of automobiles, and household size. The interview is then continued only if a household falls within a cell for which respondents are still needed (a form of 'quota sampling').

A related issue concerns the number of categories into which a researcher desires to "break down" a variable. For example, if six age groups are to be analyzed rather than two, a larger sample size will be needed to assure an adequate number of responses in each group.

Finally, imagine a study for which 500 completed surveys are necessary. One must then attempt to predict the response rate. If a 90% response rate can be expected, then only 556 respondents need be sampled; if a 50% response rate is likely, then 1000

TABLE 2   CONFIDENCE INTERVALS BY SAMPLE SIZE FOR VARIOUS PERCENTAGE SPLITS[a]

	Percentage Split		
Sample Size	50-50	75-25	90-10
50	13.9%	12.0%	8.3%
75	11.3	9.8	6.8
100	9.8	8.5	5.9
150	8.0	6.9	4.8
200	6.9	6.0	4.2
250	6.2	5.4	3.7
300	5.7	4.9	3.4

[a]For 95% confidence

		PERSONS/DU				
CROSS CLASS	AUTOS/DU	1	2	3	4	5+
SINGLE FAMILY	0					
	1					
	2+					
MULTI- FAMILY	0					
	1					
	2+					

FIGURE 8   FLORIDA STANDARD TRIP RATE MATRIX
(Schimpeler. Corradino Associates 1983). DU
= dwelling units.

must be sampled. This calculation affects the size of print runs, the amount of labor hired, and the time within which the survey will be completed. If advance letters are mailed to persons to be interviewed, and the required sample size is reached prior to interviewing everyone on the list, it may be necessary to complete the interviews anyway, if only for good public relations. The pilot study may also indicate the potential response rate.

Sampling Frames

Prior to selecting a sample, the *population* or *universe* to which inference will be made must be defined. This population may be all city residents, all households (the definition of which is often not clear — sometimes two households live under the same roof; some people live in group quarters), elderly residents, or students in a school. A list (or *sampling frame*) is needed of all of the sampling units within this population. In some cases, this is a simple task. If a sample of all students in a school is to be drawn, it is likely that a listing is available. If a sample of dwelling units in a small city is needed, aerial photographs may be available.

In many cases, however, the development of a sampling frame is a tedious process. Geographers are often interested in drawing a random sample of all residents of a city, yet no sampling frame exists for such a population. The telephone directory includes only those households with telephones who choose to be listed. Backstrom and Hursh-César (1981:113) indicated that 28% of households choose to be unlisted in the United States. Although the bias introduced by unlisted households is not fully established, those with unlisted numbers are probably less educated, younger, less likely to join voluntary associations, and more likely to be blue-collar workers, divorced, or single females (Brunner and Brunner 1971:121-124). A telephone directory includes persons who have already migrated out of the study area as well as business numbers, which are inappropriate for a survey of residences. Telephone companies do maintain a complete list of all telephone numbers but will not release it. Lists of all dwelling units with electrical service are not released by utility companies. While city directories may include those with unlisted numbers, they still exclude new residents. Moreover, in a telephone survey they do not provide a means for reaching those with unlisted numbers.

Kish (1965:53-59) presented four potential problems with sampling frames. First, many sampling frames contain missing elements. Telephone directories, in addition to omitting those without telephones and households with unlisted numbers, are also biased against new residents and may exclude portions of the study area or include unwanted territories. Lists of students in a school may exclude part-time students or those registering after the beginning of the school year. In either case, clear biases exist in the sampling frame. Solutions include redefining the population to match the sampling frame or obtaining supplementary sampling frames. Maps or old aerial photographs may not include recently-constructed features. The solution might be a rather tedious field survey and/or commissioning a new set of photographs.

Second, some sampling frames list groups of sampling units, and not individual sampling units. A list of all dwelling units is not a list of all households: two households may reside in one dwelling unit. Also, if the desired sampling unit is the individual and the sampling frame lists households, a procedure must be devised to select randomly one respondent per selected household (see below).

Third, some sampling frames contain elements which are not part of the population for which inference is to be drawn. For instance, a random sample of *full-time* students at a university would be difficult to draw from a list with no indication of student status. A list of dwelling units may contain some that have already been destroyed. If a sample of males is desired, it will probably be possible to separate most males from females by referring to first names. On the other hand, a screen question may be necessary to determine if a selected sampling unit is eligible for a survey, such as finding individuals in given age cohorts. Because most lists will probably contain some unwanted elements, it is best to oversample and then discard unwanted elements as they are selected. A common mistake is to sample the element listed directly after an unwanted element. This is incorrect. The selected element has had two chances to be selected (once for itself and once after the element just prior to it), but in practice the bias introduced by this procedure is probably not significant.

Fourth, many sampling frames contain duplicate listings. Many households are listed in the telephone directory more than once, due either to error or because they have more than one telephone number. In the latter case, such households are likely to be of higher income or run a home-based business. Sometimes duplicate listings occur because the sampling frame includes more than one list. In such a situation, it is best to cross-check the lists, producing one master sampling frame.

Babbie (1973:91) suggested that one should be careful to note any sampling frame problems when reporting results. Survey findings should be reported as representative only of the sampling elements on the sampling frame and the researcher should assess omissions or duplicate listings and note their effects.

Maps can also be used as sampling frames. If farms are being sampled, for example, a procedure can be used to randomly (or systematically) place points, traverses, or quadrants on a map. These locations can then be found on the earth's surface and treated as a random sample of the population (Yeates 1974:48-55).

One often-overlooked sampling frame is the reverse telephone directory, available for many United States urbanized areas, usually from Bresser's (1984). This company computerizes the telephone directory and publishes it in both telephone number and address order. Each listing includes the street address, name, telephone number, zip code, length of residence, and a business phone indicator. Income level, obtained from the appropriate indicated census tract, is also listed. Roseman and Williams (1980), in attempting to find new residents of an area, compared the 1970 and 1975 telephone books. Had reverse directories been available for their study area, much work would have been saved. For a study of the impact of the new Miami downtown people mover on travel and land use in the Miami CBD, a sampling was needed of all business establishments located in a defined area of downtown Miami. *Bresser's Cross-Index Directory* (1984) was used to list all business establishments on randomly-selected block faces.

McLafferty and Hall (1982) needed to conduct home interviews with persons who travelled along one particular street. To develop a sampling frame, a license plate survey was conducted by simply recording the plate numbers on cars travelling on that street and then accessing state records to obtain addresses.

Cox and McCarthy (1980) needed to interview community activists in a telephone survey. In a straight random sample, the proportion of times such a person would be reached would be too small to yield an acceptable sample size. Therefore, they subjected a local newspaper to content analysis to identify regions of the metropolitan area where locational conflicts had recently arisen. They then proceeded to over-sample in those areas.

In some instances, devising a satisfactory sampling frame could be beyond the cost and time limits of the survey. In these cases, two-phase, multistage, or snowball samples can be used (see below). Some survey mechanisms do not require a sampling frame, such as intercept and random digit dialing surveys.

The Random Digit Dialing Telephone Survey

A number of problems have been noted with telephone directories. For example, movers are more likely to be younger and better educated than nonmovers, and moves are often generated by divorce and separation. Thus, the lack of inclusion of recent arrivals may introduce a significant bias. The extent of this bias varies spatially. In larger urban areas, a greater residential mobility and a greater percentage of persons with unlisted numbers is likely. Fowler (1984:21) suggested that in some central cities, almost 50% of households are excluded from the directory. In small, relatively stable rural communities, the telephone directory may provide a satisfactory sampling frame. In some cases, however, random digit dialing (RDD) presents a far better alternative in that it provides a mechanism for reaching unlisted households. On the other hand, the procedure does not correct biases introduced when households have zero, or two or more, telephone numbers.

The first step in the implementation of an RDD telephone survey is to obtain a map of telephone exchange codes from the phone company. From this map, or *Bresser's Directory,* a listing of all exchanges may be compiled. A serious problem may be introduced if the exchanges are not coterminous with the study area, as is often the case, when the study area consists of a muncipality, a voting precinct, a school district, or any such legally-bounded space. In the Ann Arbor-Ypsilanti Transportation Study (Schimpeler.Corradino Associates 1980), 90% of the calls in one exchange would have been made outside the study area, and so the study area was redefined slightly to exclude this exchange. For another exchange, about 50% of the respondents would be from outside the study area. A screen question was devised to disqualify them. Unfortunately, respondents including those in small, poorly defined areas, such as informal neighborhoods or unincorporated areas, may not know if they live within the area defined by the interviewer. Some exchanges may also be eliminated because they are for specialized uses and contain no households. (In Dade County, the 284 exchange is reserved for the University of Miami and 579 and 638, for county government offices.) Exchanges containing mostly business numbers can also be eliminated using Bresser's Directory. (In Dade County, *Bresser's* shows that over 95% of exchange 377 are business numbers.) One must be certain that the map of exchanges is completely current; omitting an entire exchange could introduce a serious bias, particularly because exchanges are most often added in newly-settled areas.

The second step in an RDD procedure is to generate four-digit random numbers for each exchange. *Bresser's Cross-Index Directory* can be helpful in this process as well, because the range of the assigned telephone numbers can be determined for each exchange. For example, it may be that for exchange XXX, only telephone numbers into the 3000s have been assigned. In this case, generating random numbers of 4000 and above will result in numerous fruitless dialings.

A significant drawback to RDD is that many calls do not reach a residence, but are made to businesses, telephone booths, and "not in service" numbers. Much time is wasted making these calls and an expense added to the survey (Lyons and Durant 1980). Experience shows that only about 20% of dialings reach a residence with someone home (Glasser and Metzger 1972; Barrett 1983). Fowler (1984:32) reported that about 25% of possible numbers belong to residential housing units, 30% in urban areas and 10% in rural areas. He also suggested that the 'hit rate' can be increased to over 60% by conducting an initial screening of numbers within groups of 100 (Area Code-255-22XX, for example). If a residential number is found within the first 10 or so attempts, calls to that 100 series continue.

Dillman (1978:241) discussed an experiment in which respondents were identified from the telephone directory and then called. Half the calls were made asking for the person listed in the directory; half were called under RDD conditions (the interviewer did not ask to speak with the person listed in the directory). No differences were found in response rates and response quality between the two groups of interviewees. In a second experiment, respondents selected by a directory procedure were sent advance notice of the impending call; those selected as if they had been chosen by RDD were called without advance notice. The response rate for the former group was 92%; for the latter, 85%. Also, those called with advance notice provided more complete answers and a greater degree of cooperation. Therefore, Dillman suggests that a drawback to RDD is an inability to send advance notice. However, he ignored the use of *Bresser's* for finding names and addresses, given the telephone numbers. Because this directory is no more accurate than the telephone directory, if a number is not found, it still must be dialed without advance notification. In many instances, however, one can eliminate calls to businesses and obtain most names and addresses for sending advance notice. This procedure can only be used in areas covered by a reverse telephone directory, however. Another drawback is that the reverse directories are expensive to rent. They are often available from local chambers of commerce who house them under the condition that they are to be used only on an occasional basis by individuals. For a statewide or nationwide survey, coverage by reverse directories is likely to be incomplete and renting all necessary volumes would probably entail an unacceptable cost.

Thus, considerable thought should be given to the choice between RDD and directory methods. For a national sample, one would have to obtain 4700 telephone directories, making RDD superior. If population turnover is important, then RDD may be best. However, a directory method may prove effective, particularly if a survey is timed to occur soon after the directory is issued and the percentage of unlisted numbers is known not to be too great. The survey purpose may play a major role in this decision. If it is to examine the attitudes of elderly, long-term residents, the telephone directory may be effective. For a study of hazard perception among new residents, RDD may be superior. Readers interested in the RDD methodology should consult Daniel (1979).

A variation of RDD is the 'plus one' method, whereby numbers are taken from the directory and 1 is added to the last digit. This should result in contacting a satisfactory number of persons who are not listed in the telephone book (Groves and Kahn 1979).

Sample Designs I: Probability Sampling

The necessity for sampling, the issue of sample size and the development of a sampling frame have been discussed. This section deals with procedures for sampling from a population of spatially discrete phenomena, such as dwelling units, farms, or industrial plants. The discussion below focuses on probability sampling designs: simple random, systematic, stratified, cluster, multistage, and two-phase samples. Nonprobability samples (purposive, quota, and snowball sampling) are addressed in the next section.

Simple Random Sampling

In a simple random sample (SRS), each sampling unit and every possible combination of sampling units has an equal chance of selection. On a list of households, each household must be numbered and random numbers drawn to select the sample households. Such numbers should be drawn from a table which has been guaranteed to contain numbers in random order (Rand Corporation 1966). The proper way to use such a table is as follows:

(1) Treat each random number as a multi-digit decimal number. Assume that the first number in the table is 84298, now .84298.
(2) Determine the highest number on the sampling frame. Suppose it is 500.
(3) Multiply .84298 by 500, yielding 421. Thus, element 421 should be included in the sample.

Note that this procedure will always yield a number between 1 and 500 for a sampling frame of 500. If one were to follow the procedures suggested in some geography texts, three-digit random numbers would be selected from the table and any number greater than 500 would be ignored. Unfortunately, random numbers tables are not devised so that all three-digit numbers less than 500 are random after the removal of the intervening numbers which are greater than 500. Thus, the described procedure should always be used.

A technical point is that if one samples without replacement, then each subsequently chosen sampling unit has a slightly greater chance of being selected. Strictly, when drawing a sample, a selected sampling unit should be replaced and may be selected repeatedly. In actual practice, this is rarely done and if one is sampling 400 households from a list of 1,000,000, the change in the probability of each element being selected varies only very slightly and the probability of an element being selected multiple times is small. If, however, one is sampling from a relatively small population, Blalock's (1979:557-558) discussion of correction factors is useful.

Most importantly, notice that a random sample is different from a 'haphazard' sample. Finally, note that computer software also exists which can read files and randomly select some given number of elements.

Systematic Sampling

Often, sampling frames are not numbered and, particularly if there is a varying number of entries on each page (or a varying number of ineligible entries on each page, such as business listings in a phone book), the first task is to identify and number all entries in the sampling frame. This task can be very time consuming. If the ordering of a sampling frame can be assumed random, a systematic sample may be drawn which is essentially equivalent to a SRS. The procedure to select a systematic sample is as follows:

(1) If one is to select 100 units from a sampling frame of 1,000,
    divide 1000 by 100 = 10. ("10" is the "sampling interval.")
(2) From a random numbers table, select a random number
    between 1 and 9; this is called a "random start."
(3) Select every 10th element after the random start.

If the calculated sampling interval is not an integer, it will have to be rounded to the nearest whole number. If the rounding is great, a sample that is significantly larger or smaller than desired may result. An alternative is to round down and then select a second sample with a much larger sampling interval or use SRS methods. For example, suppose one needs to sample 400 units from 5000. The sampling interval is 5000/400 = 12.5. Sampling every thirteenth unit would undersample, yielding only 384; sampling every twelfth would oversample, yielding 416. In this case, it is better to sample every twelfth and then select an additional 16 by SRS methods.

An additional random element may be introduced into systematic sampling by using one random start to select sampling units from the first 100 items in the above example, a second random start for the second 100 elements, etc. Thus, one might select elements 2, 12, 22, 32 . . . 104, 114, 124 . . . , given 2 as the first random start and 4 as the second.

Systematic sampling is sometimes done by the column inch in telephone directories (as when one samples the first name in each column, or the name that appears six inches after the last one that was selected) or when the sampling units are located on index cards, as is often the case with voters' registration records. Ackerman (1975) used this type of procedure when he interviewed rural households at six kilometer intervals in a study of shopping trips in rural Argentina.

Systematic sampling can produce a biased sample if periodicity exists in the sampling frame. For example, suppose one listed all households in a condominium containing six different models and that each building contains six units, one of each model. If the sampling interval selected was six, the sample could 'hit' the same model every time. If that model, was, for example, the most expensive, the survey would likely overestimate income. In another application, every tenth house might be located on a corner, and those residents might differ from their neighbors.

Systematic sampling will also lead to a bias if the list, for example, is in ascending order by income. Depending on the selected random start, income could either be overestimated or underestimated. Lists that are in alphabetical order can generally be assumed to be in random order.

Stratified Sampling

Strata are population subgroups which, for example, might be defined demo-graphically (elderly, nonelderly) or regionally (urban, suburban, rural) and from which one might draw separate samples. Note that to sample from strata, one must have a sampling frame from which it is possible to identify the stratum to which each sampling unit belongs. A sample is sometimes stratified because separate lists are available for portions of the population, and it is less effort to draw separate samples then to combine the lists. At other times, a list may contain some relevant indication of the age, race, or geographical location of the sampling unit.

Suppose one is doing a study of university students in which an important variable is whether a student is from 'in-state' or from 'out-of-state.' If the sampling frame reveals students' origins, and it is known that 60% of the students are from 'in-state,' then one should select 60% of the sample from this group. This is called a *proportionate* stratified sample, because a comparable sampling fraction is used in each stratum, resulting in a larger sample from the larger stratum (compare this to a disproportionate stratified sample discussed below).

The major advantage of proportionate stratification is that "in almost all cir-cumstances our population estimates will be more precise as a result" (Dixon and Leach 1978a:16). One could argue that a SRS of all students should produce a sample that has 60% in-state students; the *proportionate* stratified sample, however, *ensures* that such will occur, whereas in a SRS an "unusual" sample might occasionally be produced. Yeates (1974:49) noted that in a spatial sample it is possible to completely 'miss' an entire region with a random sample. For example, with a SRS, one can randomly sample households in a city without selecting any located in the downtown area. If the survey purpose was to estimate the proportion of persons walking to work, this sample would underestimate this value (because those living downtown are probably more likely to walk to work). This problem could be solved by stratifying the sample spatially and sampling proportionately from the downtown area and the suburbs.

Thus, when using a stratified sample rather than a SRS, the confidence interval for a given sample size will be narrower for a given confidence level. If one can assume that the variation within strata is small, compared with the variation between strata, then the standard error will be smaller for a given sample size then would be obtained from a SRS because the standard error of a stratified sample is the sum of the variances of each stratum:

$$SE_s = \sqrt{\frac{\sum (n_i \ s_i^2) \ (1-(n/N))}{n^2}}$$ (4)

where: SE_s =   the standard error of a stratified sample
        n_i =   the number of cases in stratum i
        s_i =   the standard deviation in stratum i
        n =   the total sample size
        N =   the population size

If a proportion is the main consideration, then the expression **PQ** is substituted for **s** in equation (4). Blalock (1979:566) provides a geographic example of the calculation of a standard error from a stratified sample.

For certain problems, a *disproportionate* stratified sample might be beneficial. Suppose, for example, that 80% of a country's households live in Region A; 20%, in Region B. An SRS of 500 households would result in a sample of 400 in A, but of only 100 in B. This would be a satisfactory sample size for '95 and 5' in A, but not in B, where '95 and 10' would be achieved. If the desired sample size in B were 400, then disproportionate random sampling would be necessary.

Another instance in which disproportionate stratified sampling might be important is when the variation in one strata is significantly greater than in another. For example, suppose that Region **X** contains only single-family homes valued at over $100,000, but Region **Y** contains single-family homes of all values as well as apartments, condominiums, and an elderly retirement center. A larger sample might well be selected for **Y** because it contains a much more varied population.

Disproportionate stratified sampling also might be used if the cost of selecting or interviewing in one area is significantly greater than in another. A smaller sample might be selected in the more expensive area.

Theoretically, an optimum stratified sample can be devised in which the sampling fraction for each stratum is proportional to the variability (as measured by the standard deviation) in each stratum divided by the square root of the cost per element in each stratum (Stuart 1968:52). However, because deviations and costs per element are rarely available prior to selecting a sample, this procedure is rarely used.

The choice of stratification variables is often determined by availability. If a sampling frame contains both the race and the age of each sampling unit, one may very well sample from age-race categories so as to reflect the age-race breakdown of the frame. Because geographical location is usually highly correlated with demographic and other variables and because many potential sampling frames often contain locational information, it is a common choice of stratifying variable.

In a RDD survey, the only possible stratification is by geographic area (given the map of exchanges), although to the extent that demographic groups tend to cluster spatially, some stratification along demographic lines is possible. For example, in Dade County, if one wanted to disproportionately sample the elderly, this could be accomplished by generating extra four-digit random numbers for the 672, 673, and 674 exchanges (South Miami Beach). If a disproportionate number of blacks was desired, extra random numbers could be generated for exchanges 691 and 696 in Miami's Liberty City. An important point is that the oversampled elderly and blacks are those who live in elderly and black areas, and they are not a random sample of all elderly or blacks in a metropolitan area.

In an intercept survey, stratifying is difficult. It is often possible, however, to stratify the sample by appropriate selection of intercept locations. For example, if one wanted to oversample Hispanic bus riders in Dade County, then one would distribute more forms (or interview more persons) on bus routes passing through Hispanic areas. In a survey of persons at a recreation site, an oversampling of fishermen might be achieved by doing additional interviews with persons carrying fishing gear.

If disproportionate random sampling is used, then care must be taken in 'putting the sample back together' during the data analysis stage. In the hypothetical county mentioned above, suppose 400 surveys are completed in Region A and 400 in Region B for a total of 800 surveys countywide. Results may be reported separately for A and B. If, however, countywide results are desired, then the results for A need to be 'weighted up' and those for B, 'weighted down.' If a *proportionate* random sample of 800 had been selected from the two strata, then 640 (80%) would have been selected from A and 160 (20%) from B. Because we have only 400 from A, we need to weight these by a factor of 1.6 (640/400); because we have 400 from B, but a countywide sample requires only 160, we must weight each of the 400 by a factor of .4 (160/400). In the *Statistical Package for the Social Sciences* (SPSS; Nie *et al.* 1975), this would be accomplished with the following statements:

```
IF          (REGION EQ 1) WF = 1.6
IF          (REGION EQ 2) WF =  .4
WEIGHT      WF
```

This same procedure is useful even if proportionate statified sampling is used. Suppose that, although our goal in Region 1 was 640 complete interviews, only 400 were actually achieved because of an unexpectedly high nonresponse rate. If exactly 160 interviews were completed in Region 2, then the above SPSS statements, minus the middle one, could be used to 'correct' for the nonresponse in 1.

Cluster or Multistage Sampling

There are two circumstances in which cluster or multistage sampling can be a satisfactory alternative. The first is when no sampling frame of individual elements exists, but one can easily identify clusters of elements within which complete enumeration of all elements is relatively simple. For example, suppose a geographer desires to study the preferences of university students concerning the state in which each hopes to live after graduation. While it would be impossible to obtain a sampling frame of all university students in the United States, one could obtain a sampling frame of all universities and then undertake a blanket sample or a SRS within each university. Each such cluster is also called a primary sampling unit (PSU). In a blanket sample, all sampling units are included in the sample. Similarly, one might be unable to obtain a listing of all households in an area, but could make a listing of all block faces and use a blanket sample on a group of randomly-selected block faces. If one selects a random sample (rather than a blanket sample) from each PSU, then the sampling procedure is called multistage sampling, instead of cluster sampling.

A second instance in which cluster or multistage sampling can be useful is if a sampling frame of all elements exists, but interviewing at geographically-dispersed locations presents a problem. While not a consideration in mail or telephone surveys, this is a serious factor in a home interview survey. A random sample of 400 unclustered households leads to a very spatially-dispersed sample and logistical problems for the interviewers. In a cluster sample, an interviewer will need to contact perhaps five households in one small area. Thus, if one person is not at home, another potential respondent probably will be available nearby.

A significant problem is that clusters tend to be relatively homogeneous due to spatial autocorrelation. That is, persons living on the same block are more likely to be demographically similar to one another than are persons living in different areas of a city. Dixon and Leach (1978a:22) suggested that the needed sample size when using a cluster sample is 1.5 times that necessary with a SRS. Table 1 can still be used, but one must look at the row for the confidence limit that is $\frac{2}{3}$ of the one desired. For example, for a SRS, a sample of only 43 would be needed (at the 95% confidence level) for a 15% confidence interval, but a sample of 96 would be needed for a 15% interval for a cluster sample (96 being associated with 10% for a SRS). While cluster samples are efficient in the field, they come at a great cost in sample size or confidence level, particularly for studies of spatial diversity.

The tradeoff in cluster sampling is obviously between the number of clusters and cost. The fewer the clusters, the lower the cost, but the more elements per cluster. If one needed a random sample of 400, the two extreme choices would be one cluster of 400 and 400 clusters of one. Obviously, the optimal choice lies somewhere between these two. Because cluster samples incorporating few clusters increase error, one should choose as many clusters as possible, subject to budget and time constraints.

In selecting clusters for a multistage sample, the best procedure is probability proportionate-to-size (PPS) sampling. Most cluster sampling involves clusters of vastly different sizes. For example, for a survey of housing values, a random sample of city blocks might be selected. If City Block A contains 100 households and City Block B contains only 10 households, when selecting clusters randomly, A should be given ten times the chance of B to be part of the sample. When randomly selecting the same number of households from each block, each household on the large Block (A) will have only $\frac{1}{10}$ the chance of being selected as those on Block B. The result is that all 110 households have an equal chance of selection. Note, however, that sampling in this manner requires knowledge of the number of households on each block prior to drawing the sample of blocks. Often, because of the effort involved in counting households on all city blocks, such information is only gleaned *after* the blocks are selected. Therefore, weighting factors must be added upon survey completion.

For the example above, in which universities were the clusters, the number of students in each university would be known *a priori,* and PPS sampling could be applied. In some instances, a decision may be made either to include a cluster if it is significantly larger than any other, or to exclude certain clusters that contain very few sampling units, to simplify the field work. For example, for a study of attitudes of downtown employees toward Dade County's METRORAIL in which business firms were sampled and surveys administered to a blanket sample of the firms' employees, the decision was made to include all of the ten largest firms (clusters). Necessary weighting factors were then applied so that the results could be inferred to the universe of all downtown employees.

A final point is that it is possible to stratify the clusters themselves, either proportionately or disproportionately. For example, one might select a certain number of urban, suburban, and rural clusters. Such is often the case when a sampling of voting precincts is used in exit polls. Backstrom and Hursh-César (1981) provided an excellent example of a two-stage cluster sample of households, particulary with respect to procedures for choosing households on selected blocks.

Two-Phase Sampling

Two-phase sampling involves the selection of an initial, first phase sample from which a second phase sample is selected. At least three reasons may be cited for following such a procedure. First, this sampling strategy can be used if a lower level of precision is satisfactory for certain variables. United States Census methods are similar to this procedure. Certain key variables are collected from a blanket sample of the entire population (the first phase sample). Additional variables are collected from a 5% sample which, however, is not selected based on the results of the blanket sample.

A second application for two-phase sampling occurs when information is needed that is very expensive to collect. For example, a travel survey may be conducted by telephone to collect key variables from a large sample. Then, on the basis of such variables as household size, automobile ownership, and dwelling unit type, a second-phase sample might be selected. An expensive home interview survey with the distribution and collection of travel diaries may then be conducted with this smaller sample.

Two-phase sampling can also be used effectively when a study design calls for the need to locate and interview a rare population. Suppose one wishes to conduct in-depth interviews with persons who ride a bicycle to work. A quick telephone interview might be used to find such persons, who are then interviewed personally in the second phase sample.

Sample Designs II: Nonprobability Sampling

The previous sampling procedures involve a concerted effort to obtain a representative 'random' sample in which each sampling unit has an equal probability of inclusion. In this section sampling designs which do not make an attempt at randomization are discussed.

Purposive Sampling

In a purposive sample, elements are selected for specific purposes. Researchers may wish to interview "experts" in a field, rather than a random sample of the entire population. This might provide useful information for designing a questionnaire for later use with a random sample. A purposive sample was used in the pilot study for the Dade County On-Board Transit Survey. Rather than selecting a random sample of five bus routes on which to distribute pilot survey forms, routes were selected in elderly, Hispanic, black, and suburban areas. This design resulted in changes in the questionnaire and in the survey logistics. The survey administrators learned about problems involved with distributing forms to the elderly on a moving bus, for example. Bourne (1976) used a purposive sampling procedure in interviewing developers in Toronto who were principally responsible for the urban renewal process. For a discussion of methods for selecting and interviewing small numbers of purposively-selected individuals, see Feldman (1981).

Quota Sampling

In quota sampling, interviewers are instructed to interview a certain number of people falling into various demographic categories so that the totals reflect census distributions. Interviewers are given free choice as to which persons in the different categories are interviewed. While this is somewhat similar in structure to a stratified sample, the fact that the interviewers select the actual respondents implies that the results cannot be inferred to any population.

Snowball Sampling

There are some populations for which a full sampling frame might be difficult to establish. Suppose, for example, a geographer were studying the manner in which immigrants from Trinidad have adapted to the United States. Because only a very small percentage of households will contain a Trinidadian, a telephone survey would result in many wasted calls. In such a case, one might obtain the membership list of the 'Trinidad Club' and interview all these people. Respondents could be asked to provide information to locate other Trinidadians. These are then added to the list and contacted. This procedure continues until respondents are only providing the names of persons already listed.

Other Nonprobability Sampling Procedures

University professors sometimes distribute questionnaires in large lecture classes. Such a methodology is simple, inexpensive, and likely to yield a response rate close to 100%, but is not useful for drawing inferences about the students at a university as a whole.

Many surveys are conducted where interviewers simply stop people on street corners. This type of intercept survey is satisfactory for discerning the characteristics of persons passing a particular point at a particular time of day, but population characteristics may not be inferred. Ward (1975) used such a technique in studying residents of 'skid rows' in six cities.

Surveys in which respondents volunteer to participate by calling one area code 900 telephone number if they vote "yes" and another 900 number for a "no" vote, will yield interesting, but obviously biased, results. Although a majority of people watching *Saturday Night Live,* who were willing to spend the time and 50 cents for a telephone call, voted to keep "Larry, the Lobster" from being boiled to death on the air, this does not imply that the majority of Americans have similar feelings about the killing of animals for food.

Respondent Selection

A final major issue in sample design is that the above procedures most often identify households. Yet, most survey designs call for interviewing one person per household. A further procedure is needed to identify interviewees. The most common methodology is illustrated by Figure 9. Note that the manner in which a respondent is selected is to ascertain the number of adults and the number of men in the household. (Asking for the number of men is less threatening than asking for the number of women.) A table is then consulted to identify the interviewee. The selections in each cell

Since we select the respondent to be interviewed randomly, I also need to know how many persons 18 years or older live in your household, counting yourself.
(CIRCLE NUMBER OF ADULTS BELOW)

VERSION III		NUMBER OF ADULTS IN HOUSEHOLD			
		1	2	3	4+
	★	Woman	Youngest Woman	Oldest Woman	Oldest Woman
NUMBER OF	1	Man	Woman	Man	Youngest Woman
MEN	2		Youngest Man	Oldest Man	Oldest Man
IN	3			Oldest Man	Youngest Man
HOUSEHOLD	4+				Youngest Man

Does that include everyone living there at the present time, 18 years or older? and how many of them are men?
(CIRCLE NUMBER OF MEN AND FIND INTERSECTION OF ADULTS TO MEN WHICH DETERMINES THE SEX AND RELATIVE AGE OF THE RESPONDENT TO BE INTERVIEWED)

For this survey, I need to speak to _____ currently living at home in your household. Is he/she at home?

FIGURE 9   RESPONDENT SELECTION MATRIX. This is an adaptation of the Troldahl/Carter Key and is used in the format illustrated by the Policy Sciences Program at Florida State University, Tallahassee.

of this table are randomly assigned and a different version of this table is randomly assigned to each day of the week. This eliminates any bias that may result from always interviewing, for example, the youngest woman in households with two adult women (as would occur if Figure 9 was always used). A complete set of such tables may be found in Backstrom and Hursh-César (1981:95-97).

In most cases, interviewing the person who happens to answer the telephone would result in an oversampling of women, because in the United States women are more likely to answer the telephone. In some cases, for example, if the information is easy to report and is only factual in nature, any adult may be a satisfactory respondent. Fowler (1984:32) reported that in the National Health Interview Survey, the "person who knows the most about the health of the family" was interviewed.

Summary

Figure 6 identifies the major issues in selecting a sample of survey respondents. The population to which inference is to be made must first be defined. Unless random digit dialing or intercept surveys are utilized, a sampling frame must be obtained. An appropriate sample size must be determined, based upon cost and time constraints and the necessity to produce information with acceptable levels of accuracy for the population as a whole and for selected population subgroups. Finally, an appropriate sampling design must be devised. These decisions are highly interrelated. If the cost of developing a sampling frame is high, less money may be available to pay interviewers. If no sampling frame is available, some form of multistage sampling might be dictated.

Figure 6 shows that it is important to select the sample for the main survey prior to selecting the sample for the pilot. Obviously, one would not want to interview someone in both the pilot and the main survey. If the pilot study sample was selected first, then those selected for the pilot would have to be given a zero probability of selection for the main survey, meaning that every element in the population would not have an equal chance of being selected for the main survey.

The Eight Example Surveys

The major decisions that were made in designing the sampling strategies for each of the eight example surveys are presented below. Some of the sampling issues for these surveys have already been discussed above and these issues are not repeated.

The GMJF Demographic Study. It was felt *a priori* that the Jewish communities in three different regions of Dade County (North Dade, South Dade, and The Beaches) were significantly different from one another, and that a 95 and 5 accuracy level should be achieved in each area. Because the number of households in the three areas are 46,000, 26,000, and 44,000, respectively, the goal was to obtain 380 returned mail surveys from each region (strata). (Table 1 modified by equation 2.) Thus, a dispro-portionate stratified sample was selected: equal numbers of surveys were sought in three different-sized regions. Expecting about a 60% response rate on the mail survey implied the need to mail about 630 surveys in each area. Expecting about a 75% response rate on the telephone implied the need for sampling approximately 840 households in each region. Actually, 1,000 were sampled to allow for respondents who could not be reached, wrong numbers, and the like. In the analysis, weighting factors were used so that each region appeared in the countywide results in its proper proportion. The only significant predefined subgroup for which results would be required was the elderly. Because it was believed that at least one-third of Dade County Jewish households were headed by an elderly person, no special effort was made to oversample this group.

A significant problem existed in that there was no sampling frame of Jewish households available in Dade County. Thus, RDD was given serious consideration. Recall that only 20% of RDD calls result in reaching a household. Because only 15% of the population is Jewish, it would be necessary to dial the phone about 100 times to reach three Jewish households. Figuring that 75% of telephone respondents would cooperate and that 60% of these would return the mail survey implied that every 1000 dialings of the telephone would yield 13.5 returned mail surveys. Thus, the phone would have to be dialed 85,000 times to achieve 1140 mail survey responses. This methodology might have been workable if detailed knowledge of the location of Jewish households was available prior to the survey. Some demographic studies have been able to identify a small number of telephone exchanges in which a very large per-centage of a defined population is found and in which a relatively large percentage of the population meets the specified criterion. This could not be done in Dade County.

In addition, many elderly Jewish persons would feel threatened by a call inquiring about their ethnic/religious background, particularly in crime-conscious Miami. They would be likely to hang up or lie about their religion (there were about 10,000 Holocaust survivors living in Dade County in the early 1980s).

Thus, a decision was made to develop a sampling frame which costs about $3,000. Three sources of Jewish households were identified. First, the Greater Miami

Jewish Federation's computer file of about 65,000 Jewish households was used. Second, membership lists were collected from 70 local Jewish organizations and synagogues and were cross-checked with the GMJF list to yield about 72,000 households. Obviously, if the sample were drawn from this list only, it would be biased toward households having a current affiliation with the Jewish community. Thus, a list of 1230 Distinctive Jewish Names (DNJ's) was developed (Schwartz, Goldberg, for example) and used to select households from the phone book. This third source identified 44,000 households; time did not allow for checking each of these against the GMJF list. However, as the sample was drawn from the telephone directory, each name was checked against the GMJF list and discarded if it was included on the GMJF list (because this household would then have had two chances to be selected). Because there were 44,000 households in the telephone book and 26,000 on the GMJF list with a DJN (of whom 8.5% were unlisted in the Dade County Telephone Directory, based on checking a systematic sample of 400), it could be assumed that the GMJF list contained about 55% of Jewish households in the county. A systematic sample was drawn from both the GMJF list and the telephone directory. As the sample was selected, households were separated into the South Dade, North Dade, and The Beaches strata.

A final issue arose with respect to respondent selection within households. Given the nature of the questionnaire, answers were desired from either the household head or his/her spouse. The telephone survey was designed to obtain all basic demographics for both. Nevertheless, a procedure was developed so that if the last digit of the telephone number was odd, the interviewer needed to speak with the female; even, with the male. This procedure was abandoned after the first day of interviewing. Many of the elderly, in particular, felt threatened by a request to speak with their spouse, and, for the 28% who are widowed, asking to speak with a dead spouse was not a positive beginning! Thus, it was decided that the slight bias that might be introduced by speaking with too many females would be more than compensated for by the increase in response rate. (The results showed that about 56% of Dade County Jewish adults are female; 66% of survey respondents were female.) This decision is supported by recent findings by Hagan and Collier (1983), who reported achieving a higher response rate without a respondent selection procedure, without introducing bias into the results.

The Dade County On-Board Transit Survey. Given that only about 4% of Dade County residents ride buses, the only efficient procedure for locating bus passengers was on a bus! A cluster sample was used in which the first stage was a sample of bus routes, the second, of bus runs, and the third, of bus riders. The first stage was drawn as a stratified sample of bus routes. Three strata were defined: routes scheduled for changes in the near future, routes serving downtown Miami not included in the first stratum, and all other routes. For the first stage of the sample, the sampling rates were set disproportionately at 100% for each of the first two strata (where a greater level of accuracy was desired) and 25% for the third stratum. For the second stage, a random sample of weekday and weekend "bus-run days" was selected. An appropriate number of bus run days to be sampled on each route was determined on the basis of equation (1). The third stage involved a blanket survey of all riders age 12 and over.

The Southeast Michigan Regional Travel Survey. No sampling frame existed for this seven-county area, so a multistage sampling process was used. A spatially-stratified random sample of traffic analysis zones (TAZ's) was selected from within the study area. The plan was to obtain detailed plat maps showing blocks and the dwelling units on each block within each sampled zone. Because the cost for these maps would have been excessive, a random sample of maps was selected within each TAZ. Each block was numbered on each map and a random sample of blocks selected. On each selected block, 3-10 houses were chosen at random to be part of the sample. At the same time, additional dwellings were selected as a back-up sample, recognizing that some percentage of selected dwelling units would be vacant and/or demolished, potential respondents might be out of town, and others might refuse to participate. The sample size of 2600 households was based upon a '90 and 5' criteria, using equation (3). (This equation can be used because this is a multistage sample, not a cluster sample.)

The UM Travel and Parking Survey. For logistical reasons, it was easier to do a blanket survey of all faculty and staff than to sample. Only about 373 forms were needed from the 13,000 students, for a '95 and 5' accuracy level. A random sample of 70 of the 2,748 class sections was selected. Some class "sections" were eliminated, because the "class" consisted of a single student taking music lessons or an independent study. Had all professors cooperated with the request for 15 minutes at the beginning of class, and had attendance been 100%, about 1200 students would have completed forms. About 710 forms (representing 100% of those distributed) were received from students. While this sample design is technically a cluster sample, unlike clusters of households, students clustered in classes are not homogenous with respect to travel habits. Thus, the sample could be treated as if it were a SRS. In any case, the sample size was more than sufficient for a '95 and 5' accuracy level.

The Survey of Academic Computing at the University of Miami. This survey was not designed for inference to a larger population. Some of the non-computer-oriented faculty would not have been qualified to answer the questions. Therefore, with the assistance of the various deans and chairpersons, a purposive sample of experts was selected. Each respondent represented his/her discipline.

The *Miami Herald* Post-Riot Survey. Because no sampling frame of Dade County households was available, a RDD survey was designed. The sample was stratified spatially to achieve a '95 and 5' result for both countywide and Liberty City samples and a '95 and 10' result for Richmond Heights. Exchanges for the latter two neighborhoods were oversampled to provide the needed accuracy levels. A snowball sample was used to identify and personally interview a sample of Liberty City residents without telephones. No respondent selection procedure was used due to time constraints: any adult answering the telephone was interviewed. The results were later weighted so that the age-sex-race composition of the sample would match the 1980 census data for Dade County.

The Dade County Elderly Mobility Study. Because this study was only a pilot test, no great efforts were undertaken to assure randomness. Instead, groups of elderly attending functions at senior citizen centers were given self-administered questionnaires. Obviously, this resulted in selection of a sample that was clearly displaying some mobility by the mere fact of attendance at the activity center. Nevertheless, this procedure was an excellent heuristic device for understanding the special problems of surveying the elderly population.

The *Miami Review* Readership Survey. The selection of the sample in this case was relatively simple because a computerized list of all Dade County subscribers was available. A systematic sample had to be selected because the available computer system had no random number generator, and no operator was available to program a random selection. Because the list only contained about 5400 Dade County subscribers, equation (2) suggested that a sample of 360 would be sufficient for a '95 and 5' level of accuracy.

4

Questionnaire Development

The successful drafting of a questionnaire is as much an art as a science. Researchers, perhaps laboring under a deadline, sometimes hastily produce questionnaires and then wonder why low response rates and information of dubious quality are produced. Any serious questionnaire effort should evolve over at least 4-6 weeks, excluding time necessary to obtain approval of the instruments and procedures from appropriate institutional bodies. An eight-step procedure should be followed (Figure 10).

First, the research problem should be defined as precisely as possible and a "Data List" compiled of the information desired. This list is compiled by defining the hypotheses to be tested and the dependent, intervening, and independent variables for which data are needed for hypothesis testing.

Second, the final report for the survey should be outlined, including, most importantly, an anticipated "List of Tables." This defines the necessary cross-tabulations and assures the completeness of the "Data List."

Third, a preliminary questionnaire should be developed, consulting available questionnaires on similar topics. Using questions asked elsewhere can facilitate geographic comparisons, although even the slightest changes in question wording can be cause for differential responses (Schuman and Presser 1978). As the probability is low that the preliminary questionnaire will closely resemble the final questionnaire, the *only* efficient procedure is to use word processing. This facilitates the production of different versions for testing and discussion in which question wording and order can be varied.

Fourth, the preliminary questionnaire should be submitted to a panel of experts, and a series of focus groups should be convened. Group dynamics, in which each expert attempts to prove his/her ability to 'pick apart' the questionnaire, is an essential aspect of this procedure. If the author of the preliminary version does not leave the first of these meetings with at least a slightly bruised ego, then the group has not functioned properly. The questionnaire should be examined in detail to ascertain that the guidelines for questionnaire construction, presented below, have been followed and that the instrument will collect the necessary information. Also, at this point it may be useful to send the questionnaire to be critiqued by other experts in the field. This is essential *even if* one is an expert in both questionnaire development and the topic of the questionnaire. Amateurs should seek as much help as possible. If the survey has been commissioned by an organization that is paying for the results, that organization must be part of this process. *The research problem is actually defined when the final questionnaire has been drafted.*

Fifth, an 'in-house' pretest, followed by an evaluative discussion, should be

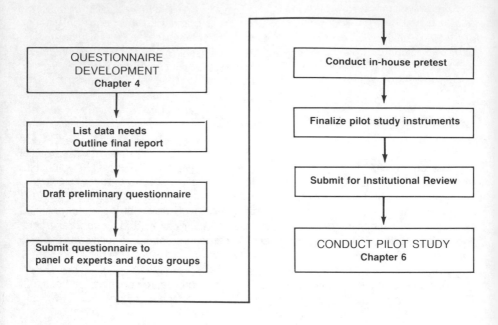

FIGURE 10   QUESTIONNAIRE DEVELOPMENT

undertaken. In a university, students are an easily available group. Unless the final sample will only include students, however, an additional group should be utilized whose demographic and occupational structure would be similar to the final sample. If the survey is of the general public, secretarial and clerical workers form a good pretest group. It may also be useful to conduct a separate pretest with any group which the panel of experts feels may experience difficulty with the questionnaire. For example, in the GMJF Demographic Study, geriatric social workers advised the survey designers that many elderly persons would experience trouble with a self-administered form. This was verified through a specially-arranged pretest and the survey mechanism was modified to allow for a telephone interview of those elderly not returning the mail survey. Whether the questionnaire is to be an interview or a self-administered form, some interviews should be completed orally. During this process, the interviewer should be particularly sensitive to the respondents' body language, which may reveal that a question is not clear, is too personal, or is one that a respondent is incapable of answering.

Sixth, at the conclusion of the in-house pretests, the questionnaire should be revised and approved for pilot testing by the panel of experts. The pilot study is a complete run through of the entire survey process and is conducted with a sample of respondents selected in the same manner as for the main survey (Chapter 6). Following the pilot study, all involved should participate in a debriefing session to discuss changes in the questionnaire or the logistical procedures.

Seventh, it may be necessary to submit the questionnaire (as well as the sampling and logistical design) to an institutional review of some kind. For example, some universities have a "human subjects" or a "survey research" committee that must approve all surveys. Such committees often do not meet frequently and sometimes as much as 4-6 weeks may be required to obtain the necessary approval. There is a trend, however, to make submission to these committees voluntary at a number of universities. That is, a researcher is only required to submit a questionnaire to this committee if possible negative psychological impact upon respondents is anticipated or the researcher is uncertain about the impact. When executing a survey under federal government sponsorship, frequently it is necessary to receive approval from the Office of Management and Budget (OMB) before proceeding. Surveys executed for local governments may have similar requirements. In Dade County, for example, the county OMB must approve all survey instruments. In any case, obtaining such approval may delay survey implementation.

Eighth, the entire questionnaire, sampling design, and survey mechanism should be re-evaluated and, if necessary, re-designed and pilot-tested again.

The closer this eight-step procedure is followed, the more likely it is that the survey will succeed. The expression 'garbage in, garbage out' is particularly applicable here. This chapter provides a series of guidelines useful for designing a questionnaire, including discussion of data types, question types, questionnaire wording, questionnaire design, and example questions.

Types of Data

While the reader is referred to Taylor (1977:39) for a complete discussion of nominal, weakly-ordered, strongly-ordered, interval, and ratio data, an understanding of data types is essential for successful questionnaire design.

Nominal data result when respondents select answers from a category or list. Examples include zip code, race, or gender. Methods used to analyze such data most often include frequency and cross-tabulation tables. The *existence* of relationships between two or more nominal variables usually is tested via a chi-square test; the *strength* of the relationship, via such measures as phi-squared and Goodman and Kruskall's Tau (Blalock 1979:279-333).

Weakly-ordered data are produced when respondents select answers from categories for which a natural order exists, such as age or income categories. Another example is a question on housing quality in which respondents reveal their perceptions of housing in a neighborhood as "luxurious, standard, or substandard." Methods of analysis are similar to those used for nominal data, except that the Kolmogorov-Smirnov test is used instead of the chi-square test (Taylor 1977:108).

Strongly-ordered data result when respondents order a set of choices. For example, one might provide a list of five cities to be ordered as to residential preference. Analytical techniques include the runs test and Spearman rank correlation (Blalock 1979:253,434).

Interval or ratio data are almost always produced by "open-ended" questions. Respondents might fill in the number of minutes spent waiting for a bus, their perception of the distance from home to work place, or their exact age. While such data are of an interval or ratio nature, respondents most often answer time and distance questions in five-minute and one-mile intervals. That is, for the question: "How many minutes did it take you to travel to work yesterday?" the vast majority will answer 5, 10, or 15 rather than 6, 16, or 18. It is difficult to argue that such data meet the 'continuously-distributed, random variable' criterion necessary for the t-test, analysis of variance, or other inferential statistics. Thus, even if an open-ended question is used, most researchers categorize the data after collection. In addition, a more difficult-to-answer, open-ended question should not be used simply because of a desire to use more complicated analysis methods.

Types of Questions

Open- and Closed-Ended Questions

In closed-ended questions, the respondent is provided a series of answers from which to choose; in an open-ended question, the respondent must fill in the answers. Open-ended questons are beneficial if:

(1) Limited knowledge exists of the responses to expect (al-though this situation can usually be ameliorated in the pre-testing and pilot surveying processes);

(2) The range of responses is liable to be great. Unless the sample size is *very* small, analysis will be difficult; any question which might elicit 200 different responses from 400 respondents probably should be reworded;

(3) One is interested in the types of responses that will be generated without prompting; and,

(4) The range of responses is great and reading or listing them would lengthen the procedure. For example, for the question, "In which county do you live?", it may be difficult to list all possible answers.

Open-ended questions have a number of serious drawbacks. First, respondents often do not read questions, or do not read them carefully. Providing answers acts as a clue to the question. For example, if all local counties can be listed without significantly lengthening the questionnaire, this is preferable to respondents answering that their "county" is the United States! Second, in an interview situation, interviewers must record responses verbatim, slowing down the interview considerably and affording the interviewee an opportunity to rethink his/her participation. This is a particular problem for telephone surveys, as long "dead spots" occur while interviewers are writing. Third, in a self-administered survey, allowing space for long, hand-written answers can lead to a lengthy questionnaire, possibly discouraging response.

Fourth, in a self-administered questionnaire, the semi-literate will find it impossible or difficult to participate. Fifth, problems arise when coders read and categorize answers written by interviewers or respondents. Much of the original flavor of the answers will be lost and, occasionally, poor handwriting causes problems. Sixth, open-ended questions require more-experienced interviewers than do closed-ended questions. Seventh, if respondents do not volunteer a particular answer, all that can be assumed is that they did not happen to think of that answer at the time of the interview. For example, if one asks, "What were the major reasons behind your move to this city?," the fact that a respondent does not mention that he/she has local relatives does not imply that this was not a major reason. A superior indicator of the importance of family in destination choice could be gained from a direct question.

Closed-ended questions avoid some logistical problems. If respondents must select among choices, at least they are categorizing themselves, and the information collected directly reflects the respondents' perceptions. A problem with closed-ended questions is that the researcher may not be able to anticipate some important dimensions. Suppose a geographer is interviewing new residents concerning perceptions of life in the Sunbelt. One approach is to ask two open-ended questions: "What do you like about Florida?" and, "What do you dislike about Florida?" A second approach is to design a question in which respondents agree or disagree with a series of statements about Florida. A problem with this approach is that factors of prime concern to many respondents may be omitted. Thus, a third approach is a combination of the first two. First, ask "What are the two things you like best about Florida?" and, "What are the two things you like least about Florida?" and then read a series of agree/disagree statements. Another approach is to ask closed-ended questions and provide respondents an opportunity to create an answer, by the inclusion of an "Other (please describe)" choice. A final approach was used by Smitt and Flaherty (1981) in a study of residents' attitudes toward exurban development. On a pilot survey, open-ended questions on the advantages and disadvantages of residential development were asked. The generated data were then used to design closed-ended questions for the main survey.

A second problem with closed-ended questions is that respondents may provide answers about topics to which they have never given any consideration. For example, suppose one of the agree/disagree statements about life in Florida concerned the quality of the local public schools. An excellent chance exists that many retirees have never reflected on this topic because it has little impact on their lives. However, they may not wish to reveal a lack of knowledge and, therefore, may provide an uninformed answer (Converse 1984).

Attitude Questions

Attitude questions obtain answers which reflect respondents' feelings about a topic. Usually, respondents are asked to answer using a "rating scale" (Edwards 1957). Such a scale may be discrete, producing weakly-ordered data. For example, one commonly-used scale is to ask if respondents "agree strongly, agree, neither agree nor disagree, disagree, or disagree strongly" with a statement. Considerable discussion exists in the literature as to whether such scales should include an odd number of choices (as the above example), which provides a neutral alternative, or an even number, forcing respondents to make a choice. Nunnally (1957:522) presents evidence that an even number of alternatives probably provides a more accurate reading of attitudes. A problem with allowing a neutral choice is that, if selected by a large number, analysis will be difficult. On the other hand, omitting this choice may force a decision from those who are truly neutral and/or do not have enough information to decide intelligently.

An alternative is to use a line upon which respondents place an "X," indicating their level of agreement or disagreement:

agree _____ **disagree**

An advantage of this procedure is that respondents can express opinions on a continuum. Moreover, the researcher obtains interval data (by measuring distance from "X" to one end of the scale). Two disadvantages are that less-educated respondents are likely to experience difficulty with this concept and an extra step (measuring) is added to the data analysis.

Respondents may also be asked attitude information using a semantic differential scale in which "polar opposites" are presented. For example, in a study of the hazard perception of hurricanes, the following scale might be used:

extremely _____**nothing to**
dangerous   **worry about**

Another method of asking attitude questions involves a cumulative or Guttman scale, which presents a series of agree/disagree statements. The statements range from very unfavorable to very favorable (on a given topic). The point at which a respondent begins to agree (after initially disagreeing) is then a measure of attitude (Backstrom and Hursh-César 1981:138).

Attitude information may also be elicited by asking respondents to rank choices from 1 to N. For example, respondents might be asked to rank five cities as to residential desirability. A problem with such a question is that respondents cannot handle a large number of choices. An alternative is the method of paired comparisons: respondents are asked to choose between City A and City B, then between B and C, then A and C, *etc.* When used with numerous pairings, this method can become rather tedious.

Note that a difference exists between preferences and attitudes, on the one hand, and choices, on the other. Although Rushton (1969) referred to "revealed preference analysis," it is actually revealed *choice* analysis. One cannot ask questions concerning current behavior and assume that it reveals preferences. If such were the case, then we would assume that most bus riders prefer buses, when in fact most would *prefer* cars, but cannot afford one.

A chief use of attitude questions in geography has been to elicit opinions about places. Gould and White's (1974) work on student views of the United States is one example, although it is unclear to what extent knowledge levels, rather than preferences and attitudes, are being measured. In other similar studies, Henderson and Voiland (1975) examined the effects of the energy shortage on residential preferences and Johnson and Brunn (1980) examined the residential preferences of blacks.

A number of other geographic studies have made extensive use of attitude questions. Examples include Gauthier and Mitchelson (1981) and Mitchelson and Gauthier (1981) in transportation research and Blommestein *et al.* (1980) and Lloyd and Jennings (1978) in consumer behavior.

Belief Questions

Belief questions measure what a respondent thinks is true or false. For example, "Do you believe that permitting the XYZ Company into the county will have a negative or positive impact?" elicits a belief about a future event. Often, however, belief and attitude questions are difficult to distinguish. Some respondents may answer "negative" to the above question because of a belief that this new industry's pollution will have a negative effect. Others may answer negatively because of a long-held attitude that 'development' is not good because of a desire to maintain a small-town atmosphere.

Attribute or Fact Questions

Fact questions gather demographic characteristics of respondents: age (Peterson 1984), sex, race, income, education, occupation (Backstrom and Hursh-César 1981:163-166), marital status, family composition, number of vehicles owned, state of previous residence, *etc.* The chief purpose for collecting such data is often to identify demographic groups which hold certain attitudes and beliefs and exhibit certain behaviors. A second purpose is to facilitate comparisons between a sample and census data so as to check the sample's representativeness (Chapter 6). For some surveys, such as a demographic survey of a group which cannot be identified via census data, these attribute variables might be the major reason for the survey. Such was the case for both the GMJF Demographic Study and for a study of urban squatter settlements by Ulack (1978).

Behavioral Questions

One of the major goals of many geographers is to describe and explain spatial behavior, particularly with respect to migration and travel behavior. This discussion concentrates on the latter.

Travel behavior surveys obtain information about where, when, how, and with how many others a respondent and household members over the age of five have traveled during a 24-hour period. A common means to elicit such information has been to ask respondents to recall all of a household's trips on a given weekday. From the early metropolitan area transportation studies, such as the Detroit Metropolitan Area Traffic Study (1956) and the Chicago Area Transportation Study (1962), to the present, most travel behavior surveys have employed this methodology. Strictly speaking, the information obtained from such a question is the respondents' *perceptions* of their behavior and the behavior of other household members. Such perceptions are likely to be flawed significantly because respondents are asked to recall a sequence of events (and details about these events) which may have seemed insignificant when they occurred. The probability of reporting errors is heightened when (as is often necessary), one household member is asked about the travel of another (termed 'proxy reporting'). Another problem is created when lengthy travel records are collected following lengthy home interview surveys and both the respondent and the interviewer are tired. Typical examples of survey instruments may be found in Stopher and Meyburg (1975) and Domencich and McFadden (1975).

A second method for collecting travel behavior information is to intercept people in the act of making a trip. The roadside interview and the on-board transit survey

(Sheskin et al. 1981) are the most common. The significant advantage of such a technique is that respondents are surveyed when they are least likely to forget trip details. For logistical reasons, however, such surveys must be kept short and it is impossible to construct a 24-hour trip record for the respondent and his/her household.

A third method is to use a travel diary on which respondents report their own behavior as they engage in travel the next day or at the end of the next day. Thus, less information should be forgotten. One problem not solved is that certain trips (such as trips hidden from a spouse) still will not be reported. Another problem is that respondents may modify behavior by postponing trips to avoid having to make entries in the diary (Dixon and Leach 1978a, 1978b). Also, in making diary entries, they may recognize and modify suboptimal behavior. Yet another drawback is that respondents may try to impress the interviewer by fabricating some trips (such as to a church).

Diary techniques have seen considerable use in television-viewing surveys and in market research (Simon 1978). Willcox (1963) found diaries to be a superior method for collecting morbidity data while Young and Willmott (1973) used them to study time budgets. McGrath and Guinn (1963) did not succeed in an attempt to combine travel diary techniques with a television show explaining how to complete the diary. In the Niagara Frontier Transportation Study, Memmott (1963) suggested that travel diaries did not produce superior results to a methodology in which every member of the household was questioned individually about past travel behavior. However, they were a significant improvement over the traditional methodology in which proxy reporting is used. Marble and colleagues (1972) and Hanson (1980) administered a five-week travel diary to 1179 households in Uppsala, Sweden. A copy of the diary form may be found in Burnett (1980). Kuzmyak and Prensky (1979) completed 285 one-month diaries with an elderly population. Other research employing travel diaries in geography includes Clark and Unwin (1980), Janelle and Goodchild (1983), Hanson and Hanson (1981), Edwards and Shaw (1982), Miller and O'Kelly (1983), and O'Kelly (1983).

Stopher and Sheskin (1982a) reviewed the above and other studies in greater depth. Four major conclusions were reached. First, convincing respondents to complete travel diairies requires a significant incentive. Second, surveillance is essential, either in the form of an appointment to pick up the travel diary (for a 24-hour diary) or repeated visits (for a long-term diary). Third, the diaries must be kept simple and explicit instructions provided. Finally, although all studies stated that the traditional methodology led to an under-reporting of trips, the evidence is mixed about the superiority of travel diaries. However, for diaries of such phenomena as consumption patterns, health problems, and expenditure patterns, Sudman and Bradburn (1974) found that even respondents who do not keep their diaries current report their behavior considerably better than if they were not keeping a diary. In two surveys completed in rural England for which about two-thirds of respondents returned diaries in a home interview situation, Edwards and Shaw (1982) demonstrated that high quality data can be achieved from travel diaries. For one of these surveys, the number of trips estimated by the survey was within +/-5% of actual traffic counts.

Figure 11 shows a travel diary designed by Stopher and Sheskin (1982a) for use in The Southeast Michigan Regional Travel Survey. The survey resulted in 2,502 complete home interview surveys with diaries from every member of the household (93% of interviewees agreed to complete the travel diaries). This multi-colored diary was designed as a booklet measuring 5 X 7 inches, so that it could be placed in a pocket or

JOURNEY NUMBER

A. WHY DID YOU GO TO THE PLACE YOU WENT TO?
CHECK ONE ONLY, PLEASE.
- ☐ HOME (GO TO C)
- ☒ WORK
- ☐ SCHOOL
- ☐ SHOPPING
- ☐ RESTAURANT/EAT MEAL
- ☐ PICK UP OR DROP OFF PASSENGER(S)
- ☐ PERSONAL BUSINESS (BANK, LAWYER, POST OFFICE, ETC)
- ☐ VISIT A FRIEND/RELATIVE
- ☐ HEALTH CARE (DOCTOR, DENTIST, HOSPITAL, ETC.)
- ☐ RECREATION (BEACH, MOVIES, THEATER, GOLF, ETC.)
- ☐ OTHER (PLEASE DESCRIBE)

B. WHAT WAS THE ADDRESS THERE?
PLEASE GIVE THE EXACT ADDRESS OR NEAREST STREET CORNER OR BUILDING NAME
For example: 18 Elm Street
OR Elm Street and Maple Street
OR The Elm Street Shopping Center

STREET/CORNER/BUILDING   2 WOODWARD AVE.

CITY/TOWN/VILLAGE   DETROIT

ZIP (IF KNOWN) |4|8|2|1|6|

C. WHAT **TRAVEL MEANS** DID YOU USE?
CHECK ONE ONLY *IF YOU USED MORE THAN ONE CHECK THE ONE USED FOR THE LONGEST DISTANCE, PLEASE!*
- ☐ WALKED
- ☐ MOTORCYCLE/MOPED
- ☐ BICYCLE
- ☒ DROVE A CAR/VAN/TRUCK
- ☐ PASSENGER IN A CAR/VAN/TRUCK
- ☐ TAXI (GO TO D)
- ☐ OTHER (PLEASE DESCRIBE) (GO TO NEXT PAGE)
- ☐ SCHOOL BUS
- ☐ REGULAR BUS
- ☐ TRAIN (GO TO F)

D. HOW MANY OTHER PEOPLE TRAVELED WITH YOU?
CHECK ONE ONLY, PLEASE.
- ☐ NONE
- ☒ ONE
- ☐ TWO
- ☐ THREE
- ☐ FOUR
- ☐ FIVE TO NINE
- ☐ TEN OR MORE

E. HOW MUCH DID PARKING COST?
2.25 ¢   ☐ NOTHING

F. HOW DID YOU GET TO THE BUS OR TRAIN?
CHECK ONE ONLY, PLEASE.
- ☐ WALKED
- ☐ BY CAR-DROPPED OFF
- ☐ BY CAR-PARKED NEARBY
- ☐ BUS
- ☐ TAXI
- ☐ MOTORCYCLE/MOPED
- ☐ BICYCLE
- ☐ OTHER (PLEASE DESCRIBE)

TRAVEL RECORD

★ TO BE FILLED OUT AS YOU TRAVEL ON YOUR TRAVEL-LOGGING DAY ★

● IF YOU DID NOT GO OUT AT ALL TODAY, PLEASE CHECK THIS BOX AND RETURN THE FORM ☐

JOURNEY NUMBER	I LEFT AT:	TO GO TO: *(Please write in the name of the place)*	I GOT THERE AT:
EXAMPLE JOURNEY	7:45 ☒ am / ☐ pm	The City-County Building	8:20 ☒ am / ☐ pm
JOURNEY 1	: ☐ am / ☐ pm		: ☐ am / ☐ pm
JOURNEY 2	: ☐ am / ☐ pm		: ☐ am / ☐ pm
JOURNEY 3	: ☐ am / ☐ pm		: ☐ am / ☐ pm
JOURNEY 4	: ☐ am / ☐ pm		: ☐ am / ☐ pm
JOURNEY 5	: ☐ am / ☐ pm		: ☐ am / ☐ pm
JOURNEY 6	: ☐ am / ☐ pm		: ☐ am / ☐ pm
JOURNEY 7	: ☐ am / ☐ pm		: ☐ am / ☐ pm
JOURNEY 8	: ☐ am / ☐ pm		: ☐ am / ☐ pm
JOURNEY 9	: ☐ am / ☐ pm		: ☐ am / ☐ pm
JOURNEY 10	: ☐ am / ☐ pm		: ☐ am / ☐ pm

EXAMPLE

10

● IF YOU MADE MORE THAN 10 JOURNEYS, PLEASE CONTINUE ON A BLANK PAGE.

FIGURE 11   TRAVEL DIARY (Courtesy of Schimpeler, Corradino Associates)

purse and carried by the respondent on the proper day. The inside front cover (left side of Figure 11) is marked for ten trips (journeys) and could be completed as a respondent travelled throughout the day. Ten pages (one for each trip) were attached on which respondents recorded additional details about each trip. The printing cost of each diary in 1980 was about $1.00, but the designers believed that the "toy" value of the diary was instrumental in persuading respondents that this task might be fun. The total cost of the entire survey effort was $124 per attitude survey with travel diary. Clearly, this procedure is expensive, although it has since been used in Honolulu at a significantly lower cost. Cost was reduced by designing a less-expensive dual survey mechanism: a telephone survey with mail-out/mail-back travel diaries on which a 65% response rate was achieved for the mail portion.

Behavioral Intent Questions

In contrast to a diary technique which attempts to elicit information on previous behavior, some surveys are designed to query about prospective behavior or behavioral intent, in which respondents predict their future behavior. "If a facility were built, would you use it?" and "Would you use METRORAIL to go to and from this campus?" are examples of such questions. Responses to behavioral intent questions are significantly less reliable indicators of future behavior than are questions about previous behavior. Couture and Dooley (1981) drew the following conclusions with respect to asking behavioral intent questions about transportation: (1) expressed behavioral intention to use transit systems overstates actual behavior; and (2) negative intent is a better indicator of nonuse than is positive intent of use. Many transportation planners agree that actual behavior can be derived from expressed behavioral intent by dividing the percentage expressing positive intentions by a number between three and five. Thus, answers to behavioral intent questions must be treated with caution.

Map Questions

Geographers may wish to use maps and/or photographs in survey research (Saarinen 1976; Downs and Stea 1973). Pacione (1983) used an interesting map technique in which thousands of individuals at meetings submitted maps of their neighborhoods. These were then used in the establishment of official districts. Unfortunately, many respondents find maps and aerial photographs to be somewhat unfamiliar and are unable to locate either their own house on a city map or their own city on a state map. Having respondents draw maps can be even trickier because one never knows if spatial knowledge or drawing skill is being tested. The use of maps in recreational surveys was discussed by the Tourism and Recreation Unit (1983) and, for environmental perception studies, by Whyte (1977). Dillman (1978:130,215) suggested that questions using maps may stimulate interest and that it is even possible to adapt some map questions to the telephone. A map on the cover of a questionnaire can highlight the local nature of a study.

Guide to Question Construction

This section concentrates on some important aspects of question wording. As implied above, question wording will change significantly throughout the questionnaire development process. Each questionnaire is unique and it is difficult to establish rules. The following guidelines are offered:

(1)   Assume that respondents will not read lengthy questions. Thus, make questions as short as possible and attempt to incorporate the question's meaning in the answers. For example, the wording of the first question below is superior.

Are you: What is your age?
• under 18 years old   • under 18
• 18-24   • 18-24

(2)   Avoid loaded words. For example, on a self-administered questionnaire, some respondents might object to a question about their "race." In areas including "Hispanics," the word "race" may be inappropriate. The question can be phrased more simply as:

Are you: • black   • white
              • Hispanic   • other _____
                                        (please describe)

(3)   Make answers mutually exclusive and exhaustive. That is, no respondent should 'fit' into more than one category and all should find an appropriate category for their situation.

(4)   Make questions as brief as possible. Unnecessary words and phrases should be eliminated. This is of particular importance on a telephone survey in which respondents will have a difficult time following lengthy questions.

(5)   Minimize the number of answers to a question. On a telephone survey, respondents cannot handle many choices. An alternative is to ask if respondents agree or disagree, and then, follow up by asking if they "strongly agree, agree, or slightly agree."

(6)   Avoid ambiguity. "How concerned are you about housing problems in this area?" is ambiguous because respondents will define "this area" differently. Similar problems are encountered in using other geographic terms, such as "suburb," "city," and "neighborhood." Belson (1981) provided an interesting discussion of ambiguous terms.

(7)   Be precise. "Near" and "far" and "frequently" and "rarely" mean different things to different respondents. Such terms should be replaced with meaningful categories of distance (under 1 mile, 1-5 miles, 6 or more miles) and time (less than once per week, once per week, more than once per week). Requesting excessive precision by using an open-ended question, however, can be unnecessarily taxing. Respondents will find it simpler to decide if their home is under or over 5 miles from a given location than to think about the exact number of miles. Thus, category-type answers are much simpler. In a similar manner, a question such as "What should the government do?" is likely to elicit responses about federal, state, and county governments.

(8)   Avoid both very simple words, which might appear condescending, and complicated words, which may be unknown.

(9)   Regional differences in langauge should be observed, such as "pop" versus "soda" versus "tonic" (Jordan and Rowntree 1982).

(10)   Avoid words that may have a specific academic meaning, but which have a different meaning to the general public. In a pilot survey of social trip behavior

(Sheskin 1974), when respondents were asked about social "trips" to visit friends and relatives, answers were given only about interurban trips. Changing the wording to social "visits" resulted in the types of answers originally expected. The question, "What is your marital status?" may evoke answers like, "As good as can be expected" and "I'll have to ask the wife."

(11)   Write questions that respondents are competent to answer. For example, "Will you ride METRORAIL to and from the University?" may evoke the response, "Where does METRORAIL run?" or even, "What is METRORAIL?." It is often necessary to design filter questions to determine respondents' level of knowledge. Asking citizens how much they think county government should spend on environmental protection will yield unusable results, because the majority of respondents have no information about county budgets.

(12)   Avoid biased questions. For example, "Some people think that redistricting is good for the state. Do you agree or disagree?" is better worded as, "Some people think redistricting is good for the state, while others think it is bad. Do you think it is good or bad?" Another example of a biased question is: "Some people think that we should devote a great deal of county resources to hurricane preparation because hurricanes can wreak havoc and destruction. Do you agree or disgree?" Similarly, avoid phrases like, "As a knowledgeable person, don't you agree?" Bias can also be introduced with a set of answers such as "agree strongly, agree, agree slightly, and disagree." Such choices should always be balanced between the agree and disagree sides. Including words such as "imperialism" or "colonialism" may prejudice a question.

(13)   Avoid linking a well-known figure or institution with a question. Asking about "Reaganomics" in the early 1980s (rather than the "current economy") was sure to yield answers reflective of attitudes toward Reagan as much as the American economy.

(14)   Only ask sensitive questions about income, religion, and politics *if absolutely necessary.* Such questions may be a source of embarrassment to respondents and can lead to ethical problems if survey workers breach confidentiality. Undue pressures should never be applied to extract answers to such questions. Some effect of the personal nature of these questions may be ameliorated by the use of broad categories (concerning income, for example).

(15)   Avoid colloquialisms and slang, as they will detract from the serious nature of the study. On the other hand, formal English is not necessary in an interview survey. Contractions, split infinitives, and ending sentences with prepositions are appropriate because interviewers will be reading the questions and should sound natural. For a self-administered form, all rules of grammar should be followed.

(16)   Avoid double-barreled questions. Asking, "Do you think that we should dismantle all 26 municipal governments in the county and consolidate all services under a central county government?" will not yield useful information. A person may say either "yes" or "no" because they agree with only one part of the question.

(17)   Avoid negative items. Respondents will often miss the word "not" and provide an answer opposite to that which they intend. Double negatives should always be avoided.

(18)   Avoid a situation in which "yes" means "no." For example, in the question, "Do you favor the repeal of the bond issue to finance public parks?" the answer

"yes" means that one is *against* expenditures for parks, while "no" means that one *favors* such expenditures.

(19) Avoid apologetic wording. "Would you mind telling me . . ." only serves to raise a flag that something personal is to follow. Do not identify a group of demographic questions as "personal questions."

(20) Avoid abbreviations. Abbreviations, such as *etc., i.e.,* and *e.g.,* are not known by everyone.

(21) Avoid pronouns in one question that refer back to another. If this style must be used, then the questions should be numbered as **1A, 1B, 1C,** so that the pronouns in **1B** and **1C** are clearly connected to the noun in **1A.**

(22) Use demographic categories that allow comparisons with census and other such data.

(23) Anchor statements with which respondents must agree or disagree at the end of a continuum (Fowler 1984:89). For example, given the statement, "The parks in this county are fair," respondents may disagree either because they believe them to be very good or poor.

(24) Write questions which will show variation in the population. A statement with which 100% of respondents agree is usually useless.

General Questionnaire Format

Question order and the general format of the questionnaire are almost as important as question wording. Most questionnaires should have three sections. The first section might contain some "warm-up" type questions; the second contains the bulk of the questionnaire; and the final section elicits basic demographic information.

The following guidelines are useful for designing the first section of a questionnaire:

(1) Every questionnaire should begin with an introduction which should: (a) briefly state the purpose of the study; (b) mention the sponsor of the study (indicating that the study is a "class project" is a death knell; instead, it should be identified with the university); (c) be as short as possible; (d) use commonplace wording; (e) not mention emotion-laden topics; (f) be serious rather than light-hearted; (g) be neutral; (h) avoid the word "survey" (the word "study" is far better); (i) provide a conservative, but realistic, estimate of how much time completion of the questionnaire entails; and (j) provide a promise of confidentiality or anonymity (see Chapter 5). In an interview survey, the interviewer should hesitate only briefly between the end of the introduction and the first question. One should assume that the respondent will cooperate, with the brief hesitation providing an opportunity for the respondent to question anything in the introduction or to refuse to participate. The whole introduction should be delivered in a firm, but polite manner. In a self-administered questionnaire, assume that respondents will not read the introduction carefully and embed no important instructions in it.

(2) The first question should be: (a) clearly related to the stated purpose of the questionnaire; (b) applicable to all respondents (lest respondents be given the impression that the questionnaire is irrelevant to them); (c) non-threatening and non-personal; (d) easily answered; (e) interesting; and, (f) neutral in nature.

(3) The first section of the questionnaire should only ask personal questions if such

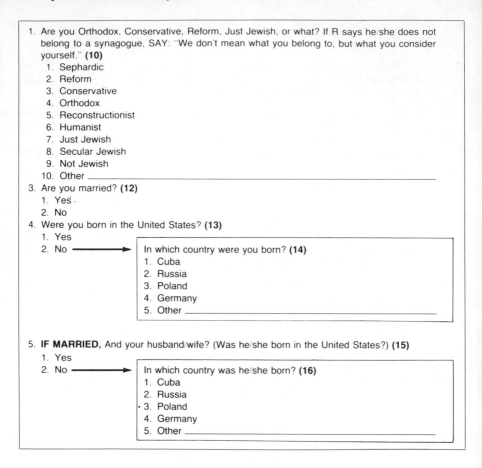

1. Are you Orthodox, Conservative, Reform, Just Jewish, or what? If R says he/she does not belong to a synagogue, SAY: "We don't mean what you belong to, but what you consider yourself." **(10)**
   1. Sephardic
   2. Reform
   3. Conservative
   4. Orthodox
   5. Reconstructionist
   6. Humanist
   7. Just Jewish
   8. Secular Jewish
   9. Not Jewish
   10. Other _____

3. Are you married? **(12)**
   1. Yes .
   2. No

4. Were you born in the United States? **(13)**
   1. Yes
   2. No ──────────→ In which country were you born? **(14)**
                      1. Cuba
                      2. Russia
                      3. Poland
                      4. Germany
                      5. Other _____

5. **IF MARRIED,** And your husband/wife? (Was he/she born in the United States?) **(15)**
   1. Yes
   2. No ──────────→ In which country was he/she born? **(16)**
                      1. Cuba
                      2. Russia
                    · 3. Poland
                      4. Germany
                      5. Other _____

FIGURE 12   THE GMJF TELEPHONE SURVEY — SOME
                  EXAMPLE QUESTIONS (Sheskin 1982)

information is necessary as a filter. For example, in The GMJF Demographic Study, it was necessary to ask if respondents were married or single because large sections of the questionnaire were not applicable to single persons. In a travel survey, it may be necessry to establish a household's level of automobile ownership if sections of the questionnaire only apply to households with cars.

(4)   In a self-administered questionnaire, the first page should be spread out and uncluttered. With a relatively long form, if respondents decide to "give it a try," and can complete the first page quickly, they are likely to continue.

Within the body of the questionnaire, the following guidelines should be used for determining question order:

(1)   Place the most important questions first. Thus, if a respondent does terminate mid-interview, much of the important information already has been obtained.

This is especially important in an interview intercept survey where respondents can easily walk away.

(2) Place those questions first which have an obvious relationship to the survey purpose and whose social relevance is more readily obvious. For example, "Should the East-West Expressway be built?" should precede "Are expressways basically good or bad for an urban area?".

(3) Place those questions first whose answers might be influenced by the 'shadow effect' of other questions. For example, the open-ended question: "What are the three most important issues facing this city?" should precede specific questions about pollution, urban sprawl, or unemployment.

(4) Group similar questions together within each content area: All "yes/no" questions should be grouped together, as should those which follow the "agree/disagree" format.

(5) Place logical content areas together. A section on the reasons for site selection for an industrial plant should be followed by the questions concerning the decision-makers. Do not skip from one topic to another!

(6) Do not place the most complex questions at the very end of the questionnaire. By then, respondents may be fatigued. Rather, a series of relatively simple questions should follow any complex question, allowing the interview to regain momentum.

(7) Place sensitive questions on each topic area toward the end of the topic section. Place all non-filter demographic questions at the end of the questionnaire. Do *not* use an "income" question as a filter. Suppose a survey concerns the crop selection decision-making process of high-income farmers. It is far better to ask questions of all farmers, with those regarding income last, and then use a cross-tabulation program to separate high- from low-income farmers.

Obviously, any questionnaire design will represent a compromise among the factors listed above.

Design Features

Questionnaire design should consider the needs of the interviewer and the coder/keypuncher for an interview survey, and the respondent and the coder/keypuncher for a self-administered form. In an interview survey, one need not pay much attention to the aesthetics of the questionnaire, whereas in a self-administered form, aesthetics could very well mean the difference between a response and a refusal. Many of the features discussed below are, therefore, important considerations in maximizing the response rate, a topic discussed in Chapter 5.

The following may be helpful in designing a questionnaire that will serve the appropriate audiences:

(1) For an interview survey, a questionnaire prepared on a word processor will facilitate changes in the development process. It provides an opportunity to "mix-up" the answers to important questions without retyping the questionnaire. The order in which choices are read can influence answers. With word processing, order can be changed easily and different versions of the questionnaire produced. A second feature of most word processors is the ability to produce boldface type. This can be used to distinguish the questions from the

---

NEIGHBORHOOD AND COMMUNITY

We would like to begin by asking some questions about your neighborhood.

1. How many **months** of the year do you usually live in **Dade** County? _____ months
If you live **elsewhere** part of the year, where is that?   _____ City
                                                                    _____ State

2. How long have you **lived** in: *Please check one in each column.*

Your Current Residence	Dade County
☐ Less than 1 year	☐ Less than 1 year
☐ 1-5 years	☐ 1-5 years
☐ 6-10 years	☐ 6-10 years
☐ 11-20 years	☐ 11-20 years
☐ More than 20 years	☐ More than 20 years

3. Why did you **move** to **Dade** County? *Please check all that you consider major reasons.*

☐ Not applicable, **born** in Dade County *(Go to Question 4)*

☐ Climate
☐ Health related
☐ Work related
☐ Retirement
☐ Large Jewish community
☐ To be near relatives
☐ To be near friends
☐ I moved from a foreign country for political reasons
☐ I came here with my parents
☐ Other: _____
          (Please describe)

4. Where did you **live before** moving to your current neighborhood? *Check one only.*

☐ A different neighborhood, but in the same area of Dade County

☐ A different area of Dade County ⟶ **Which** area was that?
                                    ☐ North Dade
☐ Outside Dade County,   ☐ South Dade
   but in Broward or Palm Beach   ☐ Miami Beach
☐ Outside Broward or Palm Beach,   ☐ Other: _____
   but in Florida   *(Please describe)*
☐ Outside Florida, but in
   the United States
☐ Outside the United States

☐ I have lived in the same neighborhood all my life

FIGURE 13   THE GMJF MAIL SURVEY — SOME EXAMPLE
                QUESTIONS (Sheskin 1982)

answers, making it simpler for interviewers to find the next question, as their eyes move from the interview form to the respondent. Instructions should be typed in all capital letters and 'boxed off.' Self-adhesive or machine lettering can also be used to enhance readability.

(2)   For a self-administered questionnaire, the advantages of typesetting and printing are enormous and more than worth the additional expense. An overall professional appearance can be given to the questionnaire; **boldface,** *italics,*

1 BELOW ARE A NUMBER OF STATEMENTS THAT MIGHT BE USED TO DESCRIBE A PERSON'S FEELINGS ABOUT MTA (METROBUS) BUS SERVICE. PLEASE TELL US HOW STRONGLY YOU AGREE OR DISAGREE WITH EACH OF THEM.

FOR EXAMPLE: SUPPOSE WE WROTE THE FOLLOWING STATEMENT:

	AGREE STRONGLY	AGREE	AGREE SLIGHTLY	DISAGREE SLIGHTLY	DISAGREE	DISAGREE STRONGLY
I LIKE THE COLOR OF THE MTA (METROBUS) BUSES.	1	(2)	3	4	5	6

SUPPOSE YOU AGREE WITH THIS STATEMENT, BUT NOT STRONGLY. YOU WOULD THEN PUT A CIRCLE AROUND THE NUMBER 2 AS SHOWN.

A NOW CIRCLE THE ANSWER CLOSEST TO YOUR FEELINGS FOR STATEMENT A.

	AGREE STRONGLY	AGREE	AGREE SLIGHTLY	DISAGREE SLIGHTLY	DISAGREE	DISAGREE STRONGLY
THINKING ABOUT METROBUS IN GENERAL, I AM VERY SATISFIED WITH THE SERVICE	1	2	3	4	5	6

NOW WE HAVE A NUMBER OF STATEMENTS ABOUT BUS SERVICE, WHICH WE WOULD LIKE YOU TO ANSWER IN A SIMILAR WAY.

	AGREE STRONGLY	AGREE	AGREE SLIGHTLY	DISAGREE SLIGHTLY	DISAGREE	DISAGREE STRONGLY
B MOST BUS DRIVERS ARE POLITE TO THE PASSENGERS	1	2	3	4	5	6
C HAVING TO WAIT TOO LONG FOR A BUS IS A MAJOR PROBLEM	1	2	3	4	5	6
D BUS SCHEDULES ARE HARD TO FIGURE OUT	1	2	3	4	5	6
E YOU CAN RELAX MORE IN A BUS THAN IN A CAR	1	2	3	4	5	6
F YOU CAN USUALLY COUNT ON BUSES TO RUN ON TIME	1	2	3	4	5	6
G WAITING FOR A BUS IN MIAMI WEATHER IS A BIG PROBLEM	1	2	3	4	5	6

FIGURE 14 THE DADE COUNTY ON-BOARD TRANSIT SURVEY — ATTITUDE QUESTIONS (Kaiser Transit Group 1982)

and various other fonts can be used to guide respondents. Lines and arrows should be professionally drawn when the typesetting is pasted-up. Typesetting will fit more on a page, without increasing clutter. Printing also permits the use of various colors and shadings to guide the respondent to the proper follow-up questions.

(3) A self-administered questionnaire should be produced on good-quality paper

and bound well. In the GMJF Demographic Study, a 5.5 by 8.5 inch booklet was produced and saddle-stitched. While this was relatively expensive, the payoff was significant. Responses to this mail survey were received from almost 80% of those to whom it was mailed.

(4)   Consider the use of colored paper to guide the interviewer or the repondent through various skip patterns.

(5)   For an interview survey, questions should be numbered consecutively throughout the form; coding numbers and computer column numbers should appear next to each question. Everything possible should be done to assist the coder/keypuncher.

(6)   For a self-administered questionnaire, questions should be numbered starting with a **1** within each topical section, making it difficult for respondents to figure the total number of questions prior to deciding to participate. No column numbers should appear. Column numbers clutter the form and remind respondents that the *COMPUTER*(!) will be used. Each answer should have a small box next to it for the repondent to check. Any type of numbering of the

---

2   A. WHERE WERE YOU **GOING TO** WHEN YOU TOOK THAT BUS RIDE? *CHECK ONLY ONE, PLEASE.*
TO:

    ☐ HOME
    ☐ WORK
    ☐ SCHOOL
    ☐ RESTAURANT
    ☐ SHOPPING
    ☐ A FRIEND/RELATIVE'S HOME
    ☐ HEALTH CARE (DOCTOR/DENTIST/HOSPITAL /ETC.)
    ☐ HOTEL/MOTEL
    ☐ AIRPORT
    ☐ RECREATION (BEACH/PARK/THEATER/MOVIES/SPORTS/ETC.)
    ☐ OTHER *(PLEASE DESCRIBE)* _____

B. WHERE IS THAT? *PLEASE TELL US THE ADDRESS, OR NEAREST STREET CORNER, OR THE BUILDING OR PLACE NAME.*

STREET/CORNER/BUILDING _____

CITY/TOWN/VILLAGE _____   ZIP |3|3| | | |
                                                                     (IF KNOWN)

C. HOW DID YOU GET TO THAT ADDRESS FROM THE LAST BUS THAT YOU USED? *CHECK ONE ONLY, PLEASE.*

    ☐ WALKED
    ☐ BY CAR – PICKED UF
    ☐ BY CAR – PARKED NEAR THE STOP
    ☐ TAXI OR JITNEY
    ☐ BICYCLE, MOTORCYCLE, OR MOPED
    ☐ OTHER *(PLEASE DESCRIBE)* _____

FIGURE 15   THE DADE COUNTY ON-BOARD TRANSIT SURVEY — BEHAVIOR QUESTIONS (Kaiser Transit Group 1982)

In this next question, I want you to think about what paying twice as much for gasoline/diesel fuel would mean. That is, suppose that the next time you filled up with gasoline/diesel fuel, you had to pay $2.70 per gallon.

5. Please look at the ORANGE response card. I am going to read a list of things that people might do because of high gasoline/diesel fuel prices, or gasoline/diesel fuel shortages. After I read each one, please tell me if . . .
    1. You started to do this regularly more than a year ago . . .
    2. You started doing this regularly within the past year . . .
    3. You would do it if gasoline/diesel fuel prices were to double next week . . .
    4. You would do it if you could buy only TEN gallons (35 liters) of gasoline/diesel fuel a week for each registered vehicle starting next week . . .
    5. You would do it either if prices doubled or if gasoline/diesel fuel were rationed . . .

	More than a year	Past Year	Double Price	Ration	Either	None	N/A	No Response
a. Observe the 55 mph speed limit	1	2	3	4	5	6	7	9
b. Take a vacation closer to home	1	2	3	4	5	6	7	9
c. Shop less frequently	1	2	3	4	5	6	7	9
d. Carpool or vanpool to work/school	1	2	3	4	5	6	7	9
e. Cancel a vacation trip	1	2	3	4	5	6	7	9
f. Combine car journeys you used to make separately	1	2	3	4	5	6	7	9
g. Buy a car that gets better mileage	1	2	3	4	5	6	7	9
h. Take the bus or train to work/school	1	2	3	4	5	6	7	9
i. Have the car tuned up regularly	1	2	3	4	5	6	7	9

FIGURE 16   THE SOUTHEAST MICHIGAN REGIONAL TRAVEL SURVEY — ATTITUDE QUESTIONS (Courtesy of Schimpeler. Corradino Associates)

answers may imply to respondents that one answer is more "important" or is a "better" choice than another. The respondents' perceptions are of primary concern, as it is they who are giving their time and can discard the form. Coding and keypunching procedures can be designed for which the lack of column and coding numbers is not a problem.

(7)   Consistency is most important. In an interview survey, the circling of the number should always be done to the left of the answer, as should the checking of a box for the self-administered form. If the answers are stacked in one column for one question, they should be stacked in one column on the others, rather than switching formats.

(8)   On both types of questionnaires, an overall uncluttered appearance is important. On a self-administered form, however, one must balance this factor against the need to present a questionnaire that appears to be brief.

(9)   For a self-administered questionnaire, it is often important to make certain that a filter question and its follow-ups appear on the same page. If a question has multiple parts, it is best to present all of them on the same page.

(10) Always provide a section for comments on the back of the questionnaire to accommodate respondents who might otherwise feel that their concerns have not been addressed.

The Example Surveys — Some Sample Questions

Although space limitations do not permit presentation of examples of all types of questions and design considerations discussed above, some of the more important points are illustrated below using questions from four of the eight example surveys.

Figure 12 shows three kinds of questions from the GMJF Demographic Study telephone questionnaire. Question 1 is closed-ended: the three most common choices are presented as part of the question, but varied answers can be circled by the interviewer. Question 3 is a personal question, normally placed toward the end of the questionnaire, but needed here as a filter for Question 5. Questions 4 and 5 illustrate filter questions with follow-up questions in boxes. Notice that while the follow-ups are written as open-ended questions, these do not present problems because the answers

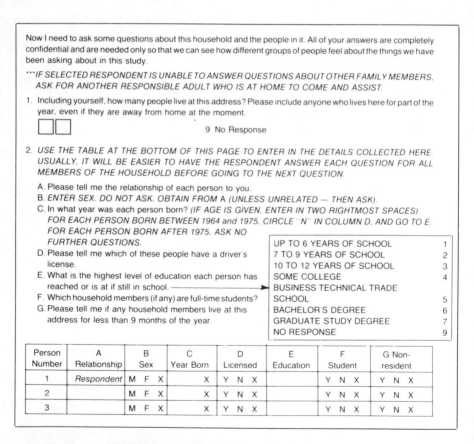

FIGURE 17   THE SOUTHEAST MICHIGAN REGIONAL TRAVEL SURVEY — DEMOGRAPHIC QUESTIONS (Courtesy of Schimpeler. Corradino Associates)

will consist of one word. For all questions, each answer has a coding number and a column number.

Figure 13 shows one page of the mail questionnaire from the GMJF Demographic Study. Notice that the heading "Neighborhood and Community" keys the respondent to the nature of the questions to follow. Question 1 is open-ended, to save space. It would have been easier for the respondent if it were closed-ended (0-3 months, 4-6 months, etc.); the designers were uncertain as to the relevant category boundaries. Also, if this question occupied any additional space, Question 4 would have been split between two pages. Asking the name of the city, in the follow-up to Question 1, resulted in useless information, because too many unique answers obviated analysis. Question 2 illustrates the use of a double-column technique to save space.

Question 3 is an attempt to make an open-ended into a closed-ended question by providing choices and permitting the respondent to check all that are considered "major" reasons. Note that a filter question ("Were you born in Dade County?") was avoided to save space and because the *a priori* feeling was that very few respondents were born in Dade County. (In fact, only 4% were!)

Question 4 produced very useful information, with the exception of the "boxed off" follow-up question. A problem arose because respondents interpreted "North Dade," "South Dade," and "Miami Beach" differently. Note that the answers are presented in the order of increasing distance from their current neighborhood.

Note as well the use of different typefaces: italics for instructions, boldface large type for major headings, slightly larger, bold numbers (so that fewer respondents would overlook questions), and boldface for the important words (so that non-readers of the questions might still see the most important words).

Figure 14 shows some attitude questions used as "warm-ups" on the Dade County On-Board Transit Survey. Instructions were provided for these respondents because they were expected, on average, to have relatively low levels of education. A decision was also made to not permit respondents to be neutral on any issue. Grouping the statements together (**B, C,** and **D** and then **E, F,** and **G**) assisted respondents in following the rows.

---

21. Suppose: • MetroRAIL opens January 1, 1984
  • The price of gas remains as it is now
  • MetroRAIL costs **$2.00** round trip ($1 per ride) plus **25¢ transfers** to and from the bus
  • Parking at a Dade County MetroRAIL station is **$1.00/day**

Would you use MetroRAIL to go to and from this campus?

  A. Never or only in an emergency   **B.** Once a month or less
  C. About 1-7 times a month   **D.** About twice a week   **E.** Every day

Would you use MetroRAIL **if instead . . .**

22. a ride costs **$1.00** round trip **A.** No   **B.** Yes
23. a ride costs **$2.50** round trip **A.** No   **B.** Yes
24. a ride costs **$3.00** round trip **A.** No   **B.** Yes
25. a ride costs **$1.00** round trip
   and **parking** is **free A.** No   **B.** Yes

FIGURE 18   THE UM TRAVEL AND PARKING SURVEY —
BEHAVIORAL INTENT QUESTIONS (Sheskin and Warburton 1983)

Figure 15 also shows some questions from the Dade County On-Board Transit Survey. Because they were all concerned with the same topic (the bus trip on which the respondent received the form), they were labeled **2A, B,** and **C.** In **2A,** the answers presented at the top of the list were expected to be the most popular. This is efficient because fewer respondents would have to read through the long list. Question **B** illustrates an effective procedure for obtaining address information. Question **C** shows a procedure for avoiding a follow-up question to the answer "car."

Figure 16 shows a question from the Southeast Michigan Regional Travel Survey. Respondents were given a response card containing the five choices. The response cards were color coded and typeset. Figure 17 shows a table-type question for collecting basic demographic information about each member of a household. This format requires significant interviewer training.

Figure 18 shows a behavioral intent question from The UM Travel and Parking Survey. Although relatively specific answers were sought for Question 21, it was felt to be too taxing to include all these choices for Questions 22-25.

Conclusion

This chapter began by stating that questionnaire design is as much an art as a science. The discussion in this chapter should convince the reader that composing a questionnaire is not a simple task. Many difficult decisions have to be made. One must balance the needs of different participants in the survey research process and try to anticipate the reactions of respondents to various questions and procedures. The most important point is that a questionnaire is not something to be prepared in a hurry. The production of a successful survey instrument is a thoughtful process requiring experienced personnel.

5

Survey Logistics

A geographer employing survey research must be familiar with the advantages and disadvantages of various survey mechanisms (Chapter 2), sampling issues (Chapter 3), and the development of survey instruments (Chapter 4). This chapter addresses an equally-important issue: survey logistics. The most technically-competent survey can 'come apart' if the logistics are not planned properly. Although the logistical *plans* for the Dade County On-Board Transit Survey were sound, the unsatisfactory quality and the quantity of the labor force hired to administer the survey, coupled with changes in administrative personnel over the two-year period during which data analysis occurred, led to survey results which were considered unreliable. As suggested in Chapter 1, such failures are not at all unusual in survey research. Thus, the researcher needs to pay significant attention to the types of issues addressed in this chapter (Figure 19).

While personal interview, telephone, mail, and intercept surveys are logistically different, a number of common issues are addressed below. These include the hiring and training of personnel, need for space to conduct the survey, optimal timing of the survey, design of control forms, survey ethics, employing methods to encourage response (including publicizing the survey and offering incentives to respondents), and budgeting. The reader is referred to Frey (1983) and Dillman (1978) for excellent discussions of the details of telephone survey logistics; to Dillman (1978) and Babbie (1973) for mail surveys; to Backstrom and Hursh-César (1981) and Warwick and Lininger (1975) for home interview surveys; to the Tourism and Recreation Unit (1983) for intercept surveys at recreational sites; and to Backstrom and Hursh-César (1981) for intercept surveys at polling places.

In general, the most logistically-complex survey is the home interview, while the mail survey is usually the least complex. Dual survey mechanisms may involve a very significant degree of coordination, as one must be concerned about the logistics of two different survey mechanisms and the manner in which they interact.

Hiring and Training of Personnel

The quality of an interview survey depends, to a great extent, upon the quality of the interviewers. For self-administered surveys, the quality of the labor is significantly less important. For mail surveys, no personal contact with the respondent ensues and only clerical labor is necessary. For intercept surveys with self-administered forms, the personnel will probably be asked questions by respondents as they distribute the

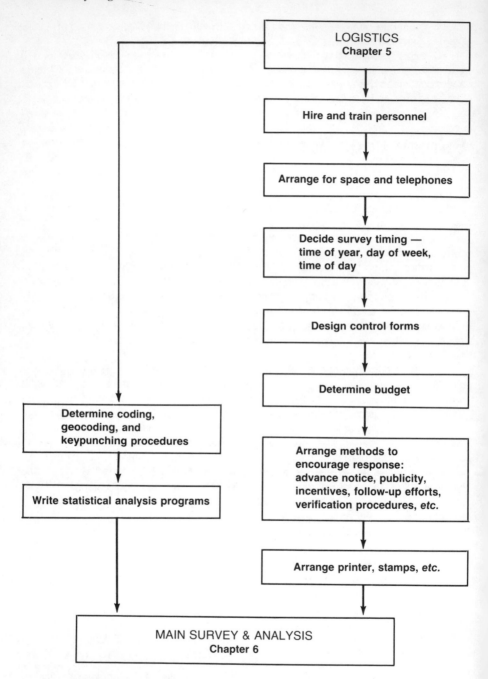

FIGURE 19   LOGISTICAL DESIGN

forms. Thus, one must be concerned about both their personal appearance and their training for answering questions about the survey.

Quality personnel must be used for home interview surveys. Workers must be well trained and trustworthy. Interestingly, such workers were found more easily twenty years ago, when many well-educated housewives were available on a part-time basis. Unlike telephone interviews, home interviews are usually not monitored. Errors, such as reading questions incorrectly or failing to select interviewees within a household according to a prescribed respondent selection procedure are difficult, if not impossible, to catch. Extensive worker training should include training sessions, practice interviews, and a comprehensive Training Manual.

Fowler (1984:108) suggested two characteristics that strong interviewers share:

(1) They have a confident assertiveness. They present the study as if there is no question that the respondent will cooperate. The tone and content of their conversation does not hint at doubt that an interview will result.

(2) They have a knack of instantly engaging people personally, so the interaction is focused on and very individually tailored to the respondent. It is not necessarily a professional interaction, but it is responsive to the individual's needs, concerns, and situation.

A third consideration is to find interviewers with these characteristics who will also not violate the norms of ethical behavior in survey research.

Potential workers should be required to attend at least two training sessions, on two different days, particularly if they are inexperienced. Some professional survey organizations train workers for as many as five days. Many potential workers will attend the first training session, but, after realizing what they are 'getting into,' fail to appear for the second. Attendance at the second session will provide a realistic estimate of the number of workers likely to appear on the first day of interviewing. At the training session, the survey administrator should present a polished introduction to the survey. An experienced interviewer should provide a demonstration interview and then workers should practice mock interviews in pairs, each playing both interviewer and interviewee roles.

When accepted for an interviewing position, workers should be given the Training Manual and instructed to read it prior to the training session, at which time the material should be reviewed. This manual should be professionally-produced and bound. While this might appear to be 'overkill,' I believe it critical to present the survey as a professional operation. This sets the "tone" for the manner in which interviewers will interact with the public.

The Training Manual should contain an introduction which explains the major survey purposes and the importance of the interviewer to the survey process. Other sections should contain explanations of the logistical procedures, samples of all survey instruments and forms, instructions for completing each form, summaries of interviewing techniques, answers to typically asked questions, summaries of the reasons for asking various questions or groups of questions, names and phone numbers of supervisors, and maps of the area to facilitate finding addresses. A number of the references cited in the Appendix, especially Backstrom and Hursh-César (1981), contain material which can be used in Training Manuals. I maintain a Training Manual on my word processor, which is modified for each survey.

For telephone interviewers, training should be just as rigorous even though, if all calls are made from a central location, interviewer errors can often be caught and remedied immediately. Although a telephone interviewer's appearance is not important, a clear, easily-understood, unaccented voice is preferable.

For some interview surveys, it can be useful to match interviewer and respondent ethnicity, while at the same time attempting to meet one's responsibility to avoid discrimination in hiring practices. Kahn and Cannell (1958) argued that if ethnicity is very important to the expected answers, then some matching can be useful. On the other hand, Weiss (1968) and the NCHSR (1977) felt that, for most survey topics, ethnicity matching does not affect data quality.

For home interviews, ethnicity matching can be accomplished by awareness of either the ethnic character of a neighborhood or the surname of the interviewee. For telephone surveys, the telephone exchange code is often a clue to respondent ethnicity. For the *Miami Herald* Post-Riot Survey, black interviewers were assigned to call Liberty City exchanges; Hispanics, Little Havana, and Hialeah; and non-Hispanic whites, middle- and upper-income suburbs. For The GMJF Demographic Study, Jewish interviewers were hired because they could relate better to the respondents, particularly the elderly, and because they needed less training in the reasons for, and the topics of, the survey.

Another issue concerns interviewer remuneration. Interviewers can be paid either by the hour or the completed interview. I have found the latter method to be advantageous in a number of surveys: workers are encouraged to be more efficient (although Dillman [1978] found no evidence to support this contention) and, because no pay is received if a respondent refuses, interviewers will hone their refusal-preventing skills. Another advantage is that the exact labor charge can be computed prior to survey execution. For the Southeast Michigan Regional Travel Survey, two different interviewing firms were engaged for the field work. One paid interviewers by the completed survey; the other, by the hour. Because of various logistical problems, the firm paying by the hour exhausted its allocation, leaving hundreds of interviews uncompleted.

Regardless of which method of payment is selected, it is important to pay workers a satisfactory wage. This will assist in attracting better-quality workers and can be effective in reducing attrition. Productivity incentives and paying for training time only if an interviewer completes some minimum number of surveys can also be effective.

Volunteer workers are almost never a bargain. Often, they are overcommitted to the 'cause' and are difficult to fire if they prove incompetent. Interviewers may be hired in a subconsultancy arrangement, professional interviewers may be hired directly, current staff may be trained, or students may be used. If students are used, it is a very good opportunity to combine a real-world and classroom experience. I have always paid students, believing this fair because the 'learning experience' ends after about ten interviews. The advantage of hiring students is that they are highly motivated, and they often have some social science knowledge. Backstrom and Hursh-César (1981:238-245) provided an interesting discussion of the advantages and disadvantages of each hiring mechanism and of the demographic characteristics desirable in interviewers.

Space to Conduct the Survey

A survey headquarters, where all operations can be centralized and personnel moved easily from one task to another is essential. The least space is needed for mail surveys; the greatest, for telephone surveys. For intercept surveys, the best location for a headquarters is in the field, as close to the interviewing site as possible. For recreational facilities, this should cause no problem, because interviewing is likely to occur at a small number of clustered locations. For the Dade County On-Board Transit Survey, questionnaires were distributed concurrently on as many as fifteen different bus routes operating in a 260-square mile area. A survey headquarters was established in downtown Miami, because so many of the routes passed through this area. The headquarters was used to store questionnaires, perform various computer coding operations, design surveyor schedules, and dispense supplies (signs, pencils, clip boards, collection boxes, survey forms). In addition, it was necessary for six supervisors to drive around the city, dealing with problems, resupplying buses with survey instruments, and shuttling surveyors from one bus to another.

A survey headquarters is critical for telephone surveys. The survey administrator may be tempted to permit interviewers to take questionnaires home, allowing them to call respondents from their personal telephones. Frey (1983:140-141) provided a number of excellent arguments for maintaining a centralized telephoning facility:

(1)   Quality control of the interviewing process is possible. Interviewers can be corrected immediately after erring and praised when they perform well.

(2)   A supervisor is always present to intervene with difficult respondents.

(3)   Coding for the computer can occur in the same room. If a coder finds something unclear on a survey form, the interviewer can be consulted soon after his/her completion of the survey.

(4)   On-the-job training is possible for interviewers hired after the training sessions.

(5)   Little opportunity exists for interview fabrication.

(6)   Weak interviewers can be identified and promptly reassigned to perform clerical tasks.

(7)   Interviewers can be assigned randomly to different exchange codes, thus controlling for any bias that may result when one interviewer calls all respondents in a region.

(8)   If the sampling design requires stratification, this can be easily monitored. For example, for the GMJF Demographic Study, 630 telephone interviews were needed in each of three regions. Because all interviewing occurred in the same room, it was possible to monitor exactly when the goal had been reached and to "move on" to the next region.

(9)   If problems arise that were not anticipated from the pilot survey, such as a question which is 'not working,' the project administrator can effect immediate change.

(10)   No completed interviews will be lost.

(11)   Fewer supervisory personnel are needed.

(12)   Interviewers who take home materials may never return.

Three other important reasons deserve mention. First, it is much easier to protect the confidentiality of the results when all interviewing and coding occur in the same location. Second, a group spirit develops when a task is performed in concert. Workers eavesdrop on one another and learn from each other. Third, workers are likely to be more diligent than if the work is taken home where distractions can interfere with the assigned task.

Thus, a centralized calling facility is imperative for a quality telephone survey. A permanent facility with individual calling carrels, push-button telephones, telephone headsets, equipment to permit the supervisor to monitor calls, and perhaps, computer terminals to permit Computer Assisted Telephone Interviewing (CATI), are optimal conditions.

Survey Timing

A number of different concerns must be considered with respect to survey timing: time of the year, day of the week, time of day, and scheduling with respect to other events in the environment. Proper timing is dependent upon the survey purpose and the geographic location of the respondents.

Summer is usually the least opportune time for a survey because of vacations, except for surveys of summer recreational facility usage. In locations with seasonal residents, surveys must be timed either to 'catch' or to miss such persons (depending upon the survey purpose). For the GMJF Demographic Study, the survey had to be conducted before April, because much of the seasonal Jewish population departs after Passover. Transportation surveys should almost always be conducted while public schools are in session because travel patterns differ at other times.

With respect to the selection of days and hours to conduct interviews, weekday evenings and weekend afternoons are generally most productive for most surveys, although it depends upon the nature of the respondent. In a survey of businesses, for example, weekday mornings and afternoons would probably be the most fruitful times. A general rule in survey research is that at least three attempts should be made to contact a respondent before 'giving up' and substituting another respondent from a back-up sample. Contact should be attempted on a weekday evening, a weekday afternoon, and on a weekend. In practice, many researchers make many more than three attempts at 'call back' in a telephone survey, because they are relatively inexpensive. In a home interview, three attempts may be the limit, particularly if a location is far from the survey headquarters and all other interviews in that area have been completed.

Severe biases can be introduced into transportation surveys, for example, by violation of callback rules. In such surveys, a major purpose is the determination of the household trip generation rate. Those households which require repeated call backs are also those which probably are involved in the greatest volume of travel! Thus, violating the callback rule would bias the sample against those who travel most. In the Chicago Area Transportation Study (1962), interviews were conducted mostly in the afternoon. Thus, the sample was biased against those households containing two working spouses (or where a single person living alone was employed), and biased toward suburban housewives. Consequently, the survey found a significant need for additional surburban highways, but underestimated the need for additional transit service.

Blair (1983) and Sudman (1980) suggested that biases have been introduced into traditional shopper surveys by the haphazard logistics usually employed in which interviewers are permitted to select respondents and in which quotas are devised for each day: interviewers commence in the morning and stop at whatever time they have reached their quota. Thus, late afternoon and evening shoppers are often missed. Blair provided evidence that this procedure has introduced significant bias into the trade area maps that result from this process.

Finally, surveys need to be timed with respect to other events in the environment. In Miami, a before-and-after impact survey was designed to gauge the effect of a new downtown people mover (DPM) on downtown travel patterns. The "before" survey needed to be timed just prior to the opening of the DPM and the "after" survey, within a few months after. This would help in attributing changes in travel to the opening of the DPM rather than to other environmental changes. As construction delays occurred, survey dates had to be revised.

Recreational surveys are particularly sensitive with respect to the day of the week, time of day, and weather conditions. At a multi-use water site, for example, usage of the facility may vary considerably on weekdays and weekends. Those arriving early in the morning may be involved in very different activities (fishing) than those arriving in early afternoon (swimming). Site use may differ from season to season as well. The number of persons visiting a zoo will vary enormously depending upon whether or not school is in session and upon weather conditions. The Tourism and Recreation Unit (1983) provided excellent examples of the scheduling of workers for intercept surveys at recreational sites.

Thus, for many types of studies conducted by geographers including transportation surveys, recreation surveys, and impact studies, timing can be a critical issue.

Design of Control Forms

Every survey generates a large volume of paperwork. For every survey mechanism, some technique must be designed to keep account of respondent contacts. For intercept surveys, counts of the number of persons passing the interviewer by time period are usually necessary. For the Dade County On-Board Transit Survey, a log sheet was used (Figure 20) on which workers recorded the time of day, the number of bus riders, and the number of persons alighting the bus at "Control Locations" spaced at about one-fourth mile intervals. The questionnaires were numbered consecutively. The numbers were used for counting the number of questionnaires distributed, by recording the number of the last questionnaire handed out at each Control Location. This knowledge was important in developing the weighting factors used for data analysis, in facilitating the calculation of a response rate, and in yielding data on ridership levels for various routes at different geographic locations.

The Telephone Control Form (TCF) used in the GMJF Demographic Study (Figure 21) illustrates the necessary records for both telephone and mail surveys. In this case, the respondent's name and address were known prior to the call and were included on the TCF to facilitate checking their accuracy with the respondent prior to sending the mail survey. Notice that space is provided to record the date and time of each call, the initials of the caller, and the outcome of the call (no answer, call back later, language barrier). Other columns provide space to record whether an interview was completed, refused, or "terminated" prior to completion; if the number was disconnected; or if an

METROPOLITAN DADE COUNTY TRANSIT SURVEY LOG SHEET
ROUTE NO. _____ DIRECTION _____ TYPE _____ BUS NO. _____ SEATS _____
NAME _____ DATE _____
START: REV. METER _____ TRANSFER METER _____
END: REV. METER _____ TRANSFER METER _____

LOCATION	PASSENGER VOLUME			TIME DEPARTED	FORM NO.		NO. REF.
	SEATED	OFF	STAN.		ENG.	SPA.	

FIGURE 20   THE DADE COUNTY ON-BOARD TRANSIT SURVEY
— LOG SHEET (Kaiser Transit Group 1982)

ineligible respondent was reached (business number, respondent not Jewish). Because this was also a mail survey, columns were provided to record the number of the mail survey (critical for computer matching of mail and telephone survey results), the date the mail survey was received, the date a second follow-up was sent (and its number), and the date a third follow-up was sent.

A control form for a home interview survey is similar in format, except that it must also include columns for noting the inability to locate a given address and for finding a given address under construction or no longer occupied.

Other forms are often necessary to keep records of personnel, the daily progress of the survey, and various other survey-specific items. The proper design of forms is often critical to the efficient control of the survey.

Coding and Keypunching

Two types of information must be precoded for computerization: address information and open-ended questions. Closed-ended questions involve no precoding — the respondent has checked the appropriate categories.

Address information must usually be precoded using a geocoding process, involving the transformation of exact address information (312 Oak Street) to a code for a census tract, zip code, traffic analysis zone, or other such geographic area. In some cities, the planning department maintains a computer program (a U.S. Census DIME file) capable of geocoding the vast majority of addresses, although such files often are dated and misspellings in the addresses keypunched from the questionnaires may prevent addresses being geocoded. In other cities, this process must be accomplished manually.

Geocoding efficiency depends on three major factors. First, the larger the region for which geocoding is to occur, the easier it will be. Second, the nature of the street

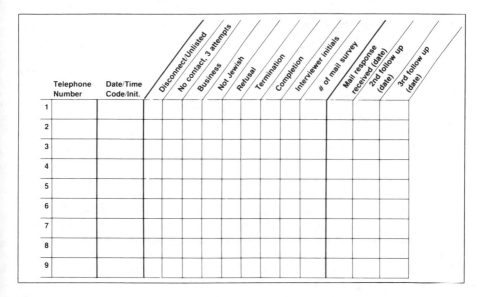

FIGURE 21   EXTRACT FROM THE GMJF DEMOGRAPHIC STUDY TELEPHONE CONTROL FORM

system in an area is critical. In Miami, for example, most roads follow a simple grid system, with streets running east-west and avenues running north-south, and with the house numbering on streets calibrated according to intersecting avenues and *vice versa*. In many cities, streets have names and no pattern to the numbering system exists. In some cases, it may even be necessary to do field work to discern house numbers on different segments of a long street. In most cities, maps and guidebooks are available that show house numbers at all intersections. In any case, geocoding is slower when streets have to be identified from an index. Third, the familiarity of the geocoders with the area involved is a significant factor. In Miami, about 50 addresses can be located in an hour by experienced personnel. In some instances, greater progress can be made, but the existence of some named streets, as well as new streets not yet on the map, often slows progress.

Precoding the answers to open-ended questions is a most difficult process because it involves a significant degree of judgment. For example, suppose respondents listed the most important problems facing their neighborhood. Each respondent is likely to give a slightly different answer. It is a challenge to create enough codes to reflect the range of answers properly without creating so many that analysis becomes impossible. A usual procedure is for a supervisor or survey administrator to create categories by coding perhaps fifty surveys. After this, the coding team should call to the attention of the supervisor any answers not easily 'fitting' under one of the existing codes; new codes may be created as one proceeds. It is always better to create more, rather than fewer codes, because categories can be easily combined, but disaggregating a category is time consuming. For example, it is probably best to create separate categories for air, water, and noise pollution and allow the computer to combine them later if warranted. This coding problem is one of the main drawbacks of the open-ended question (Chapter 3).

After geocoding and precoding, the data must be entered into the computer. The traditional procedure involved transferring the information onto 80-column coding paper or into a "coding column" directly on the form. The coding paper or questionnaires were then transmitted to a keypuncher who punched holes in computer cards. Thus, each questionnaire had to be coded and verified (checked for errors) and then keypunched and verified. Still, high error rates were not uncommon. Several far superior procedures are being used today.

The first of these procedures is Computer Assisted Telephone Interviewing (CATI) which involves placing a computer terminal in front of each telephone interviewer. The computer presents each question to the interviewer as needed (including the proper skip patterns); the interviewer types answers directly into the computer, eliminating the coding and keypunching process. If interviewers type an illegal code (such as a "3" when a question has only two answers) or an answer which is inconsistent with a previous answer (the respondent indicates possession of a driver's license and then says he is 12 years old), the computer will flash appropriate warnings. CATI also permits easy, ongoing analysis of the data throughout the survey, greatly reduces the volume of paperwork necessary to track outstanding calls, and flashes the proper telephone number at interviewers when it is time for a call back. While this procedure is technologically feasible, its use has been limited due to the time necessary to computerize the survey instrument and to bring the other necessary programs on-line and because of the significant hardware investment required for 5-10 interviewing stations (Frey 1983).

The equivalent of CATI for an intercept survey, the Ferranti MRT 100, was developed by the Questronic Project at the University of Sheffield, under the leadership of Dr. Gwyn Rowley, a Senior Lecturer in Geography (AAG *Newsletter* 1984:9; Rowley 1984). With the MRT 100, interviewers can record survey answers at the push of a button. A clipboard holds the list of questions to be asked, but the answers are fed directly into the unit. The pre-programmable MRT 100 allows up to 99 numbered multiple-choice questions per interview, and up to twelve answers per question. The question number and the answers recorded are displayed on a 16-character single-line screen. At the end of the interview, the answers can be checked and then stored in the plug-in replaceable memory unit, which can store 3,000 answers. The unit will operate for ten hours on a nine-volt built-in rechargeable battery. Answers can be fed directly into a computer for processing, eliminating the need for coding and keypunching. Interviewers can send the answers directly from the interview site, using a telephone and an acoustic coupler, or a modem can link the terminal to a survey computer.

Optical scanning (OS) forms can be used with self-administered questionnaires, given respondents who are relatively well-educated or will be given guidance in form completion. For the University of Miami Travel and Parking Survey, OS forms were distributed with the questionnaire. All faculty, administrators, students and many staff were already familiar with these forms. For those who might not be familiar with them (maintenance personnel, for example), the questionnaire was administered in groups. For some telephone surveys, the interviewer can complete the OS form as the questions are asked; in other cases, it will be easier to code answers afterward. If skip patterns are prevalent and/or many questions have more than the provided number of answers, then coding after the call is probably better. Most OS forms provide some space for alphabetic characters and groups of numbers. For example, if a respondent were asked for his/her perception of the distance between two places, such could be coded under "student number" on a typical university form. If a question has more answers than provided for on the form, two different groups of answers can be used. With a little imagination, the OS form can be used for many surveys. While it is possible to design an OS form specifically for a survey, the reprogramming of the optical scanning machine could be expensive.

Finally, 'direct keypunching' should be used if none of the above procedures are feasible. First, a coder scans each form, removing any that are too incomplete to code, making certain that answers are legible, and 'correcting' any obvious mistakes made in completing the form. If geocoding or precoding of open-ended questions is necessary, these tasks may be accomplished concurrently. Second, the questionnaires are given to data entry personnel who keypunch directly from the form. If the form is an interview survey, column numbers and code numbers will be printed directly on the form. For self-administered surveys, column numbers and coding numbers can be photocopied in the proper locations onto transparencies which can then be placed over each page of the questionnaire. (After about 50 forms, most keypunchers have memorized the codes and column numbers and no longer need the transparencies.) In any case, this procedure is far more efficient, and considerably less error-prone than the traditional procedures.

Survey Ethics

The next section addresses a number of procedures that may be used to *encourage* response. It is thus appropriate, at this juncture, to address a number of logistical issues related to the ethics of survey research. It is not uncommon for well-intentioned researchers, getting 'carried away' with the assumed importance of their particular research project, to become somewhat overzealous in attempting to encourage respondents to cooperate with the survey effort. The statistical validity of the results depends upon a high response rate. "Thus, while the researcher cannot ethically require participation, he will typically do everything possible to obtain it . . . Clearly the line between ethical persuasion and coercion is a fuzzy one" (Babbie 1973:349).

Fowler (1984:135-136) suggests that the following information be provided to each respondent at the outset of the survey:

(1)   The name of the sponsoring organization.

(2)   The name of the organization conducting the research.

(3)   The name of the interviewer (in an interview survey).

(4)   A brief description of the purpose of the survey.

(5)   A statement concerning the extent to which answers will be confidential.

(6)   Assurance that cooperation is voluntary and that respondents can skip any question they do not wish to answer.

Perhaps the most important aspect of these ethical constructs is the promise of confidentiality. It is important to distinguish between *confidentiality* and *anonymity* and to not promise the latter when it is the former which will be observed. If answers are anonymous, then the survey instruments cannot be "linked" with the respondent's name or face, even by the survey director. Such is obviously impossible in any type of personal interview survey, but can be achieved in a random digit dialing telephone survey (as long as a reverse directory is not available), a survey administered in a group setting, or in a mail survey.

Mail surveys can be made anonymous by omitting any identification number from the survey instrument. This procedure allows a claim of anonymity, which may encourage some people to respond. However, because the researcher does not know who has not yet responded, it obviates the ability to use follow-up procedures with nonrespondents only. Two possibilities exist. The first is to send reminder letters (with a second copy of the questionnaire) to all respondents. There are two problems with this procedure: (1) those who have responded are being bothered again; and (2) people who have already responded may be tempted to 'vote twice.' Similarly, perhaps one spouse filled out the first questionnaire without the other knowing; when the second questionnaire arrives, it may be completed by the second spouse without the knowledge of the first. A superior procedure for guaranteeing anonymity is to enclose, with the mail survey, a postcard with an identification number. The respondent is then requested to mail the postcard at the same time as the questionnaire. Reminder letters are then sent only to those individuals who have not yet returned a postcard. While the possibility exists that some respondents may return only the postcard, Babbie (1973) reports that the numbers of returned postcards and questionnaires are always very close.

There are occasions when research would benefit tremendously from anonymity. For example, imagine that one is surveying the opinions and economic situations of members of a club to discern the financial feasibility of building a new facility. If the researcher knows a number of club members, and if the purpose of the survey is known (which, of course, it should be), then some respondents might be reluctant to answer several of the questions. In this case, the postcard method and the guarantee of anonymity is probably the best procedure.

It is also possible to guarantee anonymity when questionnaires are administered in a group setting, such as a classroom. Note that, in a classroom-type setting, it is incumbent upon the researcher to inform potential respondents that their participation in the survey is voluntary and is not a class requirement.

Some researchers have been known to promise anonymity by claiming that there is no identification number on the questionnaire or return envelope. They use invisible ink, or a number placed under the stamp (which can be revealed after the stamp is steamed off), or a number which looks like a "form number" placed on by the printer but which is actually different on each questionnaire. Such methods are obviously unethical and should never be used.

When confidentiality is promised, the researcher must be particularly careful to protect it. Some researchers require all surveyors and coders to sign forms which promise that they will not reveal any of the information. The persons responsible for coding should be obligated to follow the same rules as the interviewers. Persons who can identify an individual from his/her demographic profile (such as a supervisor in a survey of employees) should not be permitted access to the completed questionnaires. Finally, results should not be reported for categories of people or regions which might be small enough to enable someone to discern information about individual respondents.

Respondents' names should never be written on the questionnaires; if a respondent writes his/her name on a self-administered questionnaire, it should be removed as soon as possible. The records that link respondents' names with the identification numbers should be guarded carefully and destroyed upon completion of the project. The completed questionnaires should be destroyed once the data have been entered in the computer and the researcher is absolutely certain the data are 'clean.' This can be particularly important if one wishes to protect the confidentiality of the respondents from court subpoena. Surveys are being used in courts today at an increasing rate. I recently served as an expert consultant to the FBI in a pornography trial in which a survey was commissioned by the defense to 'prove' that community standards in Miami are accepting of pornographic material. Unfortunately, the researcher made no promise of confidentiality to the respondents, requested their names, and wrote them on the questionnaires. When the questionnaires were subpoenaed, at least one respondent's confidentiality was breached when her name, her demographics, and the answers she gave to questions concerning the acceptance of materials graphically depicting sex and nudity were read aloud in open court. Even seemingly innocuous questionnaires may be subpoenaed; imagine a travel diary that was completed on a day when the respondent was accused of a crime or of cheating on a spouse. Thus, while the vast majority of surveys never become entangled in litigation, the researcher should be aware that the possibility exists.

Another important point is that some of the ethical concerns may act to influence the scientific validity of the survey. Revealing too much about the purpose of the survey may influence respondents' answers; placing too much emphasis on the confidentiality of the results may raise a flag in the minds of some respondents (suggesting that very personal questions are to be asked) resulting in a number of refusals. Nevertheless, it is incumbent upon the survey researcher to follow the rules of ethical research.

While it is important not to treat one's ethical responsibilities lightly, survey researchers generally report that most respondents react positively to the survey experience. Many are only too happy to be asked their opinions. Only a minority question the survey procedures and are suspicious of the process, including the promise of confidentiality. The main benefit to respondents for participation is intrinsic, the feeling that they have contributed to a worthwhile effort or helped to influence a decision.

The greatest breach of ethics in survey research does not involve questions of confidentiality, but rather the conduct of surveys by researchers who have not 'done their homework.' This may occur when the sample is not drawn correctly, the questionnaire is poorly designed, or the logistics are faulty. These surveys often promise respondents that the data will be used for some positive social good, but never achieve this end because of poor quality research. The development of low-cost mail and telephone surveys (versus high-cost home interview surveys) has led to a proliferation of researchers using surveys; it is incumbent upon these researchers to not waste respondents' time with surveys which will have little value.

Methods to Encourage Response

One of the most important factors in any survey is to obtain an optimal response rate (while observing the ethical concerns discussed above). Many logistical steps can be taken to encourage cooperation. Much of the previous discussion in this volume has centered on techniques for maximizing response rates; some additional factors are discussed in this section:

(1)   For home interview surveys, and particularly for telephone surveys, benefits accrue to sending postcards informing respondents of the forthcoming contact. These should be used to explain the purpose of the survey, identify its sponsor, emphasize the importance of participation, assure the respondent of confidentiality, and mention whatever incentive is to be provided (Figure 22). Postcards are more effective than letters, which need be opened and are often discarded as junk mail. The postcards should be typeset, professionally printed, and personalized. The personalization should include hand addressing and attractive postage stamps on the front. The respondent's name should be handwritten (following a printed "Dear") and the Principal Investigator should sign each postcard. All writing should be in blue ink so the respondent can see that the signature is not printed.

(2)   An advance telephone call establishing an appointment for a home interview survey can also be beneficial. Bergsten and colleagues (1984) reported a 20% cost savings with only a 1% drop in response rate following this procedure. (The advance telephone call provides an additional opportunity for refusal.) Fowler (1984:67) suggested that, when advance letters are possible, a phone survey can achieve response rates as high as those attained by personal interview surveys.

Dear

Within the next few weeks you will be receiving a phone call from the University of Miami on behalf of the Greater Miami Jewish Federation. We are doing a study of the characteristics, needs, and attitudes of the Greater Miami Jewish population to assist in planning for the diverse needs of our changing Jewish community.

We have scientifically selected a small sample of Jewish households in Dade County to receive a phone call. That is why we will greatly appreciate your participation when we call.

You may be assured of complete confidentiality. For helping us, we will be sending you (at your request) either a 2-month free subscription to the *Jewish Floridian* and/or a 1-month free membership pass to the Jewish Community Centers of South Florida.

Let me thank you in advance for your cooperation. If you have any questions, please call me at 555-1234.

Sincerely,
                                                    Dr. Ira M. Sheskin
                                                    University of Miami

---

FIGURE 22   THE ADVANCE POST CARD (Sheskin 1982:17)

(3)   Often, it is advantageous to publicize the survey. For a countrywide survey, press coverage may be possible. For a survey of some smaller target group, other means might be more beneficial. For example, a survey at a specific facility such as a club, can be publicized by posting signs or including an announcement in a newsletter distributed to the members. Such publicity can be helpful in informing potential respondents of the name of the survey's sponsor and the purpose of the survey. For a survey of the general population, such advertising is probably not cost-effective (unless it comes free): if 400 persons are interviewed in a city of one million, it is unlikely that more than a small percentage will have seen the publicity. However, publicity in a periodical aimed at a small target group can be effective. For the *Miami Review* Readership Survey, an announcement was published in a prominent location in the paper each day the survey was in progress. For the GMJF Demographic Study, announcements were placed in local Jewish publications.

(4)   Response rates can be improved by offering an incentive for respondent cooperation. This is often good public relations for the sponsor of the survey; one asks for people's time and cooperation and something is given in return. The value of the incentive is not important; most respondents are simply pleased to be offered a reward. Respondents will also feel that their opinion is considered important, because somebody is willing to 'pay' for it.

(5)   The incentive must not introduce a bias in the survey, however. For example, suppose a 10% airline discount pass was offered to respondents in a survey of vacation travel. Such an incentive would be of no value to those who never go on vacation and/or never fly. Those who do fly would be encouraged to respond. The survey then might overestimate such variables as vacation frequency and the number of times people fly in a given time period. For the Dade County On-Board Transit Survey and the Southeast Michigan Regional Travel Survey, free bus passes were offered. For the GMJF Demographic Study, both a free 2-month subscription to a newspaper and a free one-month membership in the Jewish Community Center were offered. At a minimum, respondents should be offered a copy of the survey results.

(6)   At least three attempts at telephone or home interview contact should be made for each potential respondent. On the GMJF Demograhpic Study, it was often effective for a more-skilled interviewer to call back certain respondents even after they had refused the first contact. This was not done in all instances, only in those cases in which the interviewer felt that the nonresponse resulted from a misunderstanding of the purpose of the survey or if the interviewer thought the potential respondent had been contacted at an inconvenient time. Because the survey's sponsor was a charity organization, some persons said something like "Thank you, I already gave," and hung up the telephone! While the use of more skilled interviewers in this fashion is common in survey research, it is important to remember that improving the response rate is not worth harassing the public.

(7)   For a telephone survey, interviewers will often reach answering machines. Brief messages should be left. A separate telephone number for the survey center (attached to an answering machine) should be set aside to allow respondents to return calls.

(8)   For mail surveys, white envelopes should be used for both the mail-out and mail-back (not manila envelopes, which are so often indicative of junk mail). Both envelopes should contain the words "FIRST CLASS MAIL" and at-

tractive adhesive postage stamps, preferably colorful commemorative stamps. The fact that one has already invested in the postage can be effective in eliciting response. Researchers are often tempted to use bulk mail to send out the questionnaires and business reply for the return envelopes, a false economy. The outer envelope should not identify the contents, particularly if there has been no prior telephone call to notify the respondent of the survey.

(9)   For mail surveys, the envelopes should be hand addressed, the inside cover letter dated by hand, the name of the respondent filled in, and the letter signed in blue ink (Figure 23). A cover letter from an important person or sponsor may be helpful.

(10)   For mail surveys, a complete follow-up procedure (Figure 5) should be used. This includes a thank you/reminder postcard, a second copy of the questionnaire, a third copy, and perhaps a reminder phone call. Experience shows that a second copy may increase the response rate by 50%, and a third, by another 25%. Thus, if an initial mailing achieves a 30% return; a second mailing may

---

Dear

Thank you for participating in our telephone survey and for agreeing to receive, fill out, and return the enclosed form.

This study is sponsored by the Greater Miami Jewish Federation, and its purpose is to provide the Federation and other local Jewish organizations and synagogues with information to assist in planning for the diverse needs of our rapidly-expanding and changing Jewish community.

As a token of our appreciation for agreeing to complete our questionnaire, we will send you (at your request) a 2-month introductory subscription to the *Jewish Floridian* (if you do *not* currently subscribe) and/or a 1-month free membership in the Jewish Community Centers of South Florida (if you are not currently a member) (either North Dade, South Dade, or Miami Beach). Please indicate your choice(s) on the outside of the return envelope.

You have been chosen to participate using a scientifically random procedure. It is very important that each and every household so chosen completes and returns the questionnaire so that the results will truly represent the thinking of our community.

You may be assured of complete confidentiality. The questionnaire has an identification number for mailing purposes only and so that we may send you either the *Floridian* and/or your membership in the JCC. Your name will never be placed on the questionnaire.

The results of the research will be made available to all Jewish organizations and synagogues. If you would like a summary of the results, please write, "copy of results requested" on the back of the return envelope.

We would be happy to answer any questions you might have. Please call 555-1234 and ask for Dr. Sheskin.

Thank you for your assistance.

Sincerely,

Dr. Ira M. Sheskin
Principal Investigator
University of Miami

FIGURE 23   THE COVER LETTER (Sheskin 1982:26)

yield another 15%; a third mailing, 7.5%. Spector and colleagues (1976), using a reminder letter followed by a reminder telephone call, achieved a 79% response rate to a mail survey in Sweden.

(11)   For home interview or intercept surveys, interviewers should be dressed in a respectable manner, consistent with the likely socioeconomic status of the interviewees. Interviewers should wear badges and carry letters of introduction on official stationery. Much attention should be given in training sessions concerning techniques for getting 'one's foot in the door.' For an intercept survey, professionally-designed signs should be placed at the intercept point.

(12)   A procedure should be used in which a sample of each interviewer's work is verified via a call to respondents (Chapter 6). This telephone call is also useful to gain respondent reaction to the quality of the interview. While most respondents will be reluctant to cause trouble, if an interviewer was impolite or unprofessional in any way, respondents may volunteer this information.

A significant degree of nonresponse is a serious problem in any survey, particularly if it is believed that nonrespondents differ from respondents on important survey issues. Often, nonrespondents tend to be less-educated and of lower income. From a geographic standpoint, it is likely that, because socioeconomic status varies spatially, nonresponse bias will affect areas differently. Thus, a survey may achieve a 90% response rate in an upper-income suburb, but only a 50% response rate in a poor inner-city neighborhood. The differences in the response rates must be stressed in any spatial analysis. It is much more likely that inner-city nonrespondents differ more significantly from inner-city respondents, than do upper-income suburban nonrespondents from upper-income suburban respondents.

All ethical steps should be taken in survey efforts to avoid nonresponse bias. However, if nonresponse bias does occur, several possible remedies can be applied during the analysis stage (Chapter 6).

Budgeting

One of the most difficult tasks is to devise a survey budget. On any survey, unforeseen problems may arise, particularly with respect to a lower-than-expected response rate, a higher-than-expected number of long distance calls, or a sampling frame unexpectedly found to be incomplete. Thus, it is best to estimate some factors (such as worker productivity) conservatively and hope that underbudgeting in one area will be balanced by overbudgeting in another. The following types of costs must be considered:

(1)   **Interviewer Costs.** As mentioned above, labor costs depend upon whether wages are paid by the hour or the completed survey. If paid by the completed survey, then labor costs are derived by simple multiplication; if by the hour, the number of surveys that will be completed per hour must be estimated. For self-administered intercept surveys, the number of surveys distributed is a function of the number of potential respondents passing the interviewer during different time periods. It may well be useful to monitor the survey locations prior to budgeting to establish this number. For interview-type intercept

surveys, the completion rate will depend on the survey length and the response rate, in addition to the number of persons passing by. For home interview surveys, most researchers estimate 2-4 completions per evening. For telephone surveys, completions per hour will decline as the survey progresses and more time is spent on callbacks. Particularly if random digit dialing is used, much time is spent dialing 'bad' numbers, waiting on hold, arranging callbacks, and getting busy signals. Thus, for most telephone interviews of less than 10-15 minutes, the number of completions per hour is not strongly related to length; most interviewer time is spent accessing respondents.

(2)   **Other labor costs.** A major cost can be incurred by coding and keypunching, depending upon which method is used and the volume of geocoding and the number of open-ended questions. Clerical workers, needed for creating the sampling frame, selecting the sample, photocopying, mailing, and record keeping, add to the labor cost, as do the cartographers responsible for graphics in the final report.

(3)   **Professional staff.** The greatest single expense for any survey is likely to be the salary of the professional who is responsible for designing the survey and producing the report, his/her assistants, and consultants.

(4)   **Printing Costs.** For self-administered surveys, printing costs can be significant because the survey form optimally should be typeset and printed. For telephone or home interview surveys, the form can be word processed and photocopied at a relatively minor cost. Printing costs also may be incurred for any advance notification or copies of the results sent to respondents, and for report production.

(5)   **Mailing Costs.** These costs will be incurred for either mail or interview surveys for which advance notification is used. For mail surveys, one must make certain that the questionnaire's size and weight do not exceed the projected per-unit mailing cost, keeping in mind that the envelope mailed out will contain the questionnaire, cover letter, and return envelope. Mailing costs for reminder letters may also be significant. Additional mailing costs may result if an incentive or a copy of the results is to be sent.

(6)   **Travel Costs.** For home interview surveys, interviewers and supervisors may be reimbursed for local mileage.

(7)   **Telephone Charges.** For telephone surveys, it may be necessary to install a bank of telephones or a long-distance calling service on existing telephones. While telephone usage will be less for other survey types, there may still be some long-distance calls involved with verification of home interview surveys. In 1981 dollars, commercial survey agencies charged $13 to $15 per hour for WATS (Wide Area Telephone) survey interviewing (Backstrom and Hursh-César 1981:115). A telephone answering machine or a dedicated line is also useful.

(8)   **Overhead Costs.** These costs depend on the policies of the institution through which the survey is conducted.

(9)   **Clerical supplies.** These include badges, signs, clipboards, notebooks and other such items.

(10)   **Computer time.**

(11)   **Costs for the incentive.**

(12)   **Publicity costs.**

Total costs for different types of surveys are very difficult to estimate. Dillman (1978:71) suggested that the cost for a mail survey (excluding time of the professional and overhead) in 1978 was $1.60-$2.84 per completed questionnaire. For a statewide telephone survey, he suggested that similar costs are about $7.00 per completed questionnaire, over half of which is for long-distance charges. Dillman provided evidence that a national home interview sample would cost $100 per case (again, in 1978 dollars, but including all costs), $20 per case for telephone interviews, and $6 per case for mail questionnaires. The GMJF Demographic Study cost about $40 per completed telephone/mail survey and about $30 per completed telephone survey, including all costs. The *Miami Review* survey cost $25 per completed survey because the sampling was simpler and only a telephone survey was performed. The Dade County On-Board Transit Survey attained 15,000 responses at about $15 per response. For the Southeast Michigan Regional Travel Survey, the total cost was $124 per attitude survey with a complete set of travel diaries. Thus, costs can vary significantly from survey to survey. See Backstrom and Hursh-César (1981:42) for a sample budget.

Survey Research in Non-United States Settings

This brief volume has concentrated on the science and art of survey research in North America. Geographers involved in survey research in other areas are referred to Dixon and Leach (1984) for a discussion of *Survey Research in Underdeveloped Countries,* to the United Nations (1971), Warwick and Lininger (1975), and Hursh-César and Roy (1976). Dixon and Leach (1978b) also provided a volume on survey research oriented toward the United Kingdom. While much of the foregoing is applicable to foreign situations, it is interesting to note that the only viable survey technique in most developing countries is a home interview-type survey because postal systems are unreliable, illiteracy is high, and telephone systems are inadequate. In many European situations, a telephone survey would not be successful because the cultural values shaping telephone usage differ from those of the United States and Canada.

6

Survey Execution and Data Analysis

Prior to survey execution, the survey administrator needs to coordinate three major, concurrent activities: sample design, questionnaire construction, and the design of survey logistics (Figure 3). All tasks inherent in these three activities must be completed by the date scheduled for worker training (except for mail surveys). This chapter focuses on the remaining steps: the pilot study, execution of the main survey, and data analysis (Figure 24).

The Pilot Study

A pilot study is a "walkthrough" of the entire survey procedure under real-world conditions, using a much smaller sample (perhaps 30-50 respondents) than will be employed for the "main survey." This sample may be purposive rather than random, specifically selecting respondents from particular strata. Three important purposes are served by a pilot.

The major purpose of a pilot survey is to test the sampling procedures, questionnaire design, and survey logistics. Some researchers are tempted to forego the pilot study because of its impact upon both the budget and timetable. This can be a significant mistake. Frequently, the pilot survey suggests changes which result in a significantly higher-quality final product. Sometimes these changes are extensive enough to warrant a second pilot study. I find it most useful to conduct some pilot survey interviews personally, providing direct experience with the questionnaire. The pilot survey should be followed by a debriefing session including all those who participated in it.

A second useful purpose of a pilot study is to compare two or more proposed procedures or questions which have both advantages and disadvantages (Sheskin and Stopher 1982). For example, the Dade County On-Board Transit Survey needed to query respondents' perceptions of the times and costs of alternative modes of transportation. Two alternative forms were pilot tested: a 'long form,' in which individual questions about times and costs were asked for each of three different modes, and a 'matrix form,' in which the modes formed the rows, and the time and cost parameters, the columns. A major disadvantage of the long form was that the question occupied three and one-half pages; a disadvantage of the matrix form was that the survey designers suspected that many respondents would not be able to comprehend the format of the question. An advantage of the matrix form was that it occupied less than one page. The pilot study revealed that only a slightly greater response rate could be expected from the matrix form (which was four pages shorter!) and that a large number

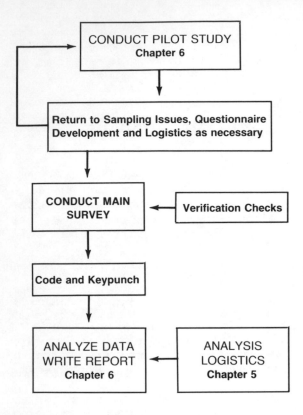

FIGURE 24   SURVEY AND ANALYSES

of respondents would experience difficulty completing the matrix. Thus, the long form was used in the main survey. Had no pilot study been done, and had the matrix form been employed instead, the results would have been seriously flawed.

As a second example, the Southeast Michigan Regional Travel Survey consisted of an attitude survey of one randomly-selected respondent per household and travel diaries for each household member over the age of four. Two possible procedures existed for performing the survey. The first procedure involved distributing the travel diaries first and then making an appointment to collect them, at which time the attitude survey would be administered (travel diaries first, interview after). The second procedure consisted of administering the attitude survey, distributing the travel diaries, and then making an appointment to collect them (interview first, travel diaries after).

Because the attitude survey was of limited utility if the travel diaries were not completed (and a high percentage of refusals to complete the travel diaries was expected), an important advantage of the first procedure was that time would not be spent on the attitude survey unless the travel diaries were completed. The second procedure, however, would permit some rapport to develop between the interviewer and interviewee during the one-hour attitude survey. It might then be easier to convince household members to complete the travel diaries. In a pilot study of 97 households, half were administered each procedure. The second procedure was clearly superior: when presented with travel diaries first, 53% refused to complete them, compared with a 4% refusal rate when preceded by the interview.

An important third purpose of a pilot test is to ascertain the length of time needed to complete the questionnaire and to obtain some indication of the likely response rate for the main survey.

Execution of the Main Survey

As a result of the pilot, procedures are finalized for the main survey. If the pilot led to no significant changes, and it was conducted on a random sample of respondents, it may be possible to count these toward the necessary sample size for the main survey. Problems sometimes arise during the main survey which did not surface during the pilot, and procedural adjustments must be made by the survey administrators. Any such changes and their effects should be noted in the final report. If the preparations for the main survey have been complete, however, then this stage of the process, while often hectic and requiring an intensive time commitment from the survey administrator, should run smoothly.

Home interview surveys involve an extra step during the main survey. A procedure should be used in which a sample of each interviewer's work is verified via a call to respondents. This call should ask if the interview did indeed occur and should re-query one or two non-threatening questions from the middle and end of the questionnaire. For the Southeast Michigan Regional Travel Survey, an interviewer was found to have fabricated more than ten interviews; this is not an isolated event.

Data Analysis

As data collection commences from the main survey, geocoding, precoding, and keypunching may begin. The statistical analysis program, originally written for the pilot, can be revised to accommodate any changes in the questionnaire. Data analysis may then begin.

A discussion of statistical techniques available to analyze survey data is beyond the scope of this volume. Readers are referred to any good statistics text such as Blalock (1979). The following discussion is a guide for the analysis of survey data.

The *first step* in analyzing survey data is to check its validity. The first computer run should produce simple frequencies of the answers to each question; unlikely or impossible answers should be checked. For example, if a value of **1** has been coded for males and **2** for females, one must ascertain why **3**'s, **4**'s, and **9**'s, for example, appear in the data. This is most often due to keypunching errors. Sometimes such errors are indicative of extensive mistakes, as when a keypuncher has punched an entire row of numbers one column too far to the left or right. To investigate the problem, one must often consult the questionnaires themselves and compare them with the computer record.

At times, values will appear which do not seem reasonable. For example, in examining answers to a question concerning the distance one must travel from home to work, suppose that 399 answers are values of "under 20 miles," and one, of 200 miles. This answer should be double checked against the questionnaire. Cross-tabulations of such items as age and driver's license possession should also be used as a data check, to make certain no computer record indicates that a 9-year old has a driver's license!

The *second step* is to check for interviewer bias: whether different interviewers elicited significantly different responses to attitude questions. This might be acceptable if the interviewers worked in demographically different areas of the study region and the differences reflect true regional variation. If the interviewees' demographic profiles are similar and one interviewer always found agreement with a statement, while another always found disagreement, then one might suspect that one (or both) of the interviewers did not follow procedures properly. Checking for such bias can also be significant if there is reason to believe that different types of interviewers may elicit varying responses. For example, in a question on race relations, white respondents may give differing answers to black and white interviewers.

The *third step* is to produce, if possible, a basic demographic profile of respondents (including gender, age, income, length of residence, geographic location, etc.) and to compare it with census data for the study area. In some cases, such as when sampling business executives, one would not expect to find a strong correspondence, and indeed, here the differences may be of interest. However, if the goal of the sampling design was to obtain a representative sample of all residents of a region, it then becomes necessary to perform a series of one sample t-tests for significant differences between sample and population means, one sample Z-tests for significant differences between sample and population proportions, and one sample chi-square or Kolmogorov-Smirnov tests for significant differences between, for example, age distributions. If all relevant tests result in a determination of "no significant difference," then one may assume that a reasonably representative sample has been achieved.

However, if significant differences are found, one of two procedures may be appropriate. The first is to assume that the census data are outdated (in the second half of the decade one may be comparing survey data with a decennial census) and that the demographic profile developed by the survey is a reasonable approximation of current conditions. The second procedure involves applying weights to the sample data so that the respondents' demographic profile reflects that of the population as revealed by the census. The procedure for such an adjustment is analogous to that discussed in Chapter 3 for adjusting for a disproportionate stratified sampling design. The choice between these procedures depends upon the extent of the demographic differences, the time lapse between the survey and the census, and knowledge concerning the extent to which the real-world environment has changed since the census year.

Suppose, for example, the census reveals that 20% of the population of a study area resides in Region A, but only 10% of the survey respondents reside in A. One procedure is to assume, on documentable grounds, that the percentage of persons in A has actually decreased since the census. The second procedure involves applying a weight of 2 to each survey completed by a resident of A. This assumes that either the sampling procedure was faulty in identifying households in A or that residents of A were more likely to be nonrespondents. One might be able to check the latter proposition, because the one piece of information usually known about nonrespondents is their approximate home address. If, however, the survey reveals 10% in an area are elderly, and the census indicates 20%, less guidance is available as to the cause of the discrepancy, because the age of nonrespondents is probably not known. In practice, this weighting procedure can be significantly more complicated if the results are to be corrected for a number of demographic/geographic variables.

The *fourth step* in the data analysis stage is to examine the possible effects of nonresponse bias. Obviously, the lower the response rate, the greater the likelihood of nonresponse bias. Aspects of this problem have already been discussed in previous chapters. Four procedures for handling nonresponse may be cited (Stopher and Sheskin 1982b).

First, a random sample of nonrespondents can be selected at the survey's completion and special, persistent efforts made to gain some brief information, while remaining within the bounds of ethical practice. For example, it might be beneficial to attempt a brief home interview with mail survey nonrespondents. This does, however, add significant cost and time to the survey.

Second, early respondents (to a mail survey) can be compared with late respondents (Donald 1960). The assumption is that late respondents are more similar to nonrespondents than are early respondents. The contention is that if no differences are found, it is less likely that nonresponse bias exists.

Third, Cochran (1963) suggested a procedure which assumes extreme values for nonrespondents. Unfortunately, the calculated confidence intervals are usually far too wide to permit meaningful inference. Instances exist, however, where this procedure can be useful. For example, suppose a survey's major purpose is to discern whether a majority of residents support construction of an expressway. Suppose a survey of 400 residents is completed with an 80% response rate, 75% (300) indicating support. If one assumes that all 100 nonrespondents would have expressed nonsupport, then, of the 500 potential respondents, 60% would have indicated support (with a confidence interval of $+/-$ 5%). Thus, the major question — do a majority support construction? — is answered affirmatively.

A final procedure for handling nonresponse bias is to plan a dual survey mechanism, including one survey mechanism for which a high response rate is expected (as in the GMJF Demographic Study and the Dade County On-Board Survey). Suppose a telephone survey that achieves a 95% response rate is followed by a mail survey with a 60% response rate. One can probably assume that little if any nonresponse bias exists in the telephone survey. Using procedures analogous to those described above for comparing survey data with census data, one could use one-sample tests to compare the demographic characteristics of telephone and mail survey respondents to ascertain if the latter are a random sample of the former. If so, then one can assume that no significant nonresponse bias has entered the mail survey. If, however, differences are found, then the mail survey results must be weighted (Stopher and Sheskin 1982b).

The *fifth step* in data analysis is to make decisions concerning 'item nonresponse.' This occurs when respondents refuse to answer a question, do not know the answer, or (in a self-administered questionnaire) overlook a question. For a question with a significant nonresponse, the decision concerning the inclusion or exclusion of non-responses is an important one. Note as an example, that in Table 3, in excluding nonrespondents, one is assuming that their income distribution is not significantly different from that of respondents. This may be an unwarranted assumption; a number of studies have shown that very-high and very-low income respondents are less likely to answer this question. In cases where nonresponse is significant, the preferred procedure is to report results both including and excluding the nonresponses.

TABLE 3   ITEM NONRESPONSE — EFFECT ON CALCULATED PERCENTAGES

Income Group	Percentage Including "No Response"	Percentage Excluding "No Response"
Under $25,000	20.0	28.6
$25,000-50,000	40.0	57.1
Over $50,000	10.0	14.3
No response	30.0	
Total	100.0	100.0

It may also be useful to use discriminant analysis to estimate missing values. For example, suppose that of 400 respondents, 50 have refused to answer the income question. Discriminant analysis can be used to develop a model which can successfully predict income for the 350 respondents who provided income information using such variables as educational level, residence location, and housing type. Once developed, this model can then be used to estimate missing values for the 50 respondents not answering the income question. While the values provided may not be a perfect reflection of reality, this procedure is far superior to simply assuming that the income levels of nonrespondents are distributed in the same manner as are those of respondents.

The *sixth step* in data analysis is to run the statistical programs. Because the majority of data collected on most questionnaires are on a nominal scale, the most important statistical procedures are also among the simplest. While multivariate statistical procedures can be applied to some types of survey data, in actual practice much survey analysis involves relatively simple techniques.

Most data analysis efforts involve the production of two types of tables: simple frequency tables, showing the number and/or percentage of respondents selecting a particular choice for each question, and cross-tabulations, in which matrix-type tables show the number and percentage of respondents who answered question A in a given manner and also answered question B in a given manner.

Other useful statistical programs include analysis of questions for which multiple answers may be indicated by each respondent; statistical descriptions which produce means and standard deviations for interval- or ratio-scaled data; and routines to produce means and standard deviations for an interval- or ratio-scaled variable for various nominal categories. An application of the latter would be the production of mean journey to work distances for various occupational categories.

Reporting the Results

The final step in survey research is to report the survey results (Figure 24). Backstrom and Hursh-César (1981:381-409) provide some excellent suggestions for survey reporting writing. While report writing is beyond the scope of the current volume, several points should be noted.

First, a full explanation of all survey procedures and logistics should be provided in either a separate chapter or an appendix. This should include descriptions of the sampling frame, sampling procedures, the response rate, methods used to encourage response, the characteristics, training, and supervision of the interviewers, the questionnaire design process, the exact wording of all questions, and the quality control procedures used during coding and data entry.

Second, one must clearly state any biases arising from the sampling strategy, problems in survey execution, the nature of the questions, or the type of population surveyed. Even if significant flaws developed in procedures or questions, the results may still be useful if the reader is appraised of these problems (Alwin 1978a and 1978b).

Third, survey results can be easily misinterpreted. In particular, one must clearly report percentages so that readers understand the denominator. I have found that many people have difficulty comprehending the statement, "67% of the elderly live in the north, but only 33% of the population in the north is elderly."

Fourth, the use of word processing is critical to a quality product. Most reports will contain numerous tables, many with similar formats. Word processing can be used to replicate formats throughout a report as well as to produce various types of summaries of results by study area subregion. In a final report for the GMJF Demographic Study, a 15-page section containing 36 summary statements about each of five subregions was produced in under one hour. (The statements were typed once and then simply duplicated for each subregion, changing the subregion title and the percentages, as necessary.) This section permitted readers to gain a perspective on the subregions without the need to review 36 different tables. The tables were useful to readers interested in comparing results between the subregions. Thus, the nature of survey results lend themselves particularly well to the utilization of word processing. In addition, graphics packages available on most microcomputers for the production of bar charts and pie graphs can add significantly to the final presentation. Often, they can be critical in conveying impressions that may not be obvious from the examination of columns of numbers in tables.

A final consideration in report writing concerns the fact that many surveys are sponsored by organizations which believe the results to be proprietary. Thus, it may not be possible to disseminate survey results in the usual academic/scientific manner. At a minimum, the sponsoring organization will request that the researcher submit all publications to it for prior approval. In other cases, the researcher may be prohibited from using any of the results in published works. In either case, the academic researcher should obtain a written agreement concerning publication prior to undertaking the survey.

An additional ethical consideration is suggested by the previous paragraph. It is not unusual for an organization to commission a survey to support its position. The academic must decide whether to participate in such an effort, particularly if the intention is to disseminate the results if, and only if, the desired result is achieved. When faced with such a situation, I have maintained the position that I will undertake an unbiased survey effort, the results of which will be reported fully to the sponsoring organization, regardless of whether they support the organization's position. If the organization then chooses to report results selectively, the researcher cannot be accused of bias. If, however, a researcher feels that the results may be seriously abused and used in an unethical manner by the sponsoring organization, then one is obliged to decline participation in the project.

Conclusion

This volume began with the premise that a survey effort should be undertaken only when no alternative methodology existed for satisfactorily examining a particular research problem. The information to be gained from a survey must be significant enough to justify asking respondents to participate in the survey process. This latter ethical consideration should always be weighed heavily in a decision to undertake a survey. Many researchers attempt to collect data via surveys without having had any training in such research. As should now be clear, successful surveying of human subjects is a time consuming, costly, and painstaking process. It involves expertise in selecting an appropriate survey mechanism, survey sampling, questionnaire construction, development of logistical procedures, project management, data and word processing, and knowledge of statistical analysis. Social scientists in general, and geographers in particular, must recognize that survey research is a powerful technique when properly applied, but can lead to improper conclusions when poorly applied. This volume should serve as a guide to the effective utilization of survey research methodologies. At the same time, it has introduced complex topics that can be pursued in some of the excellent books listed in the appendix and the bibliography. The greatest teacher, however, is experience.

APPENDIX

A number of excellent textbooks on survey research should be added to the library of readers who plan serious endeavors utilizing survey research. Each has its own merits.

Babbie, E. R. 1973. *Survey Research Methods.* Belmont, CA: Wadsworth Publishing Company. 384 pp.

Although somewhat dated, this text, intended for undergraduates taking their first course in survey research methods, provides a practical introduction, including discussion of the scientific context of survey research, survey designs (cross-sectional, longitudinal, *etc.*), survey sampling, questionnaire design, and mail and interview survey logistics. While the section on data processing technology is dated, the discussion of pretests and pilot tests is particularly well done. The volume also includes significant sections on analysis of survey data and the ethics and uses of survey research. The author is a sociologist.

Backstrom, C. H. and G. Hursh-César. 1985. *Survey Research, Second Edition.* New York: John Wiley. 436 pp.

This update of the original 1963 classical work by the same title emphasizes the home interview in presenting a "Model Survey" which serves as an example throughout the text. Brief sections on panel studies and intercept, telephone, and mail surveys are found at the end of chapters on sampling, question and questionnaire design, interviewing techniques, and data processing. The most useful feature of this text is the inclusion of 36 "Checklists" covering such topics as steps in survey research, sources of housing unit data, locating respondents, demographic items, design schematics, interviewer conduct, probing, steps in data processing, and tips for writing reports. The authors are political scientists.

Dillman, D. A. 1978. *Mail and Telephone Surveys, The Total Design Method.* New York: John Wiley. 325 pp.

This is the most complete and useful volume on mail and telephone surveys. The Total Design Method (TDM) relies on a theoretical view of the reasons people do and do not respond to questionnaires. The basic tenet is that every small administrative detail that might affect the quality and quantity of response must receive attention. Each respondent is made to feel that a personal effort is being made to elicit his/her opinion. Dillman achieves response rates of 75% for mail surveys of the general public and 80-90% for telephone surveys. The procedures of TDM have been, for the most part, incorporated in this Resource Publication. Dillman discusses the need to examine alternatives to the traditional face-to-face interview survey, the advantages and disadvantages of mail, telephone, and face-to-face surveys, and provides a detailed discussion of all of the various steps necessary for writing and implementing mail and telephone surveys. No researcher involved in a mail or telephone survey should proceed without examining this volume. The author is a sociologist.

Dixon, C. J. and B. E. Leach. 1978a. *Sampling Methods for Geographical Research.* Norwich, UK: Geo Abstracts. 47 pp.

Dixon, C. J. and B. E. Leach. 1978b. *Questionnaires and Interviews in Geographical Research.* Norwich, UK: Geo Abstracts. 51 pp.

Dixon, C. J. and B. E. Leach. 1984. *Survey Research in Underdeveloped Countries.* Norwich, UK: Geo Abstracts. 78 pp.

These three brief volumes are the only previous significant works by geographers examining survey research methods. The first examines the basics of sampling; the second covers self-administered and interview techniques and has brief sections on the use of maps and photographs, diaries, and attitude and perception questions. No discussion of telephone surveys is included because this technique is considered inappropriate in a British setting. The third volume concentrates on the home interview survey in underdeveloped countries.

Fowler F. J., Jr. 1984. *Survey Research Methods.* Beverly Hills, CA: Sage Publications. 159 pp.

This recent book provides a general overview of survey research, with particular emphasis given to methods for reducing nonresponse for the various survey mechanisms, survey error, and the ethics of survey research. The author is a social psychologist with a special interest in public health.

Frey, J. H. 1983. *Survey Research by Telephone.* Beverly Hills, CA: Sage Publications. 208 pp.

As implied by the title, this volume deals almost exclusively with telephone survey methods. A very interesting chapter provides background on the social norms of telephone usage in the United States. The most valuable chapter deals with the logistics of telephone interviewing. The author is a sociologist.

Warwick, D. P. and C. A. Lininger. 1975. *The Sample Survey: Theory and Practice.* New York: McGraw-Hill. 344 pp.

This volume provides a general overview of survey research techniques, oriented toward face-to-face interview surveys. Because it was originally developed for use in Peru, many of the examples reflect an international perspective. The authors are sociologists who were associated with the University of Michigan's Institute for Social Research.

Bibliography

AAG *Newsletter*. April 1984. Washington, DC: Association of American Geographers.
Ackerman, W. V. 1975. "Development Strategy for Cuyo, Argentina," *Annals,* Association of American Geographers 65:36-47.
Alwin, D. F. (editor). 1978a. *Survey Design and Analysis, Current Issues.* Beverly Hills CA: Sage Publications.
Alwin, D. F. 1978b. "Making Errors in Surveys: An Overview," pp. 7-26 in Alwin 1978a.
Babbie, E. R. 1973. *Survey Research Methods.* Belmont, CA: Wadsworth Publishing Company.
Backstrom, C. H. and G. Hursh. 1963. *Survey Research.* New York: John Wiley.
Backstrom, C. H. and G. Hursh-César. 1981. *Survey Research, Second Edition.* New York: John Wiley.
Barrett, S. G. 1983. "Providing Public Transit to an Emerging Minicity: North Belt, Harris County, Texas," unpublished Master's thesis, Department of Geography, University of Miami.
Baumann, D. D. and J. H. Sims. 1979. "Flood Insurance: Some Determinants of Adoption," *Economic Geography* 55:189-196.
Belson, W. A. 1981. *The Design and Understanding of Survey Questions.* Aldershot, England: Gower.
Bem, D. J. 1972. "Self-Perception Theory" in L. Berkowitz (editor), *Advances in Experimental Social Psychology.* New York: Academic Press.
Bergsten, J. W. et al. 1984. "Effect of an Advanced Telephone Call in a Personal Interview Survey," *Public Opinion Quarterly* 48:650-657.
Blair, E. 1983. "Sampling Issues in Trade Area Maps Drawn From Shopper Surveys," *Journal of Marketing* 47:98-106.
Blalock, H. M. 1979. *Social Statistics, Revised Second Edition.* New York: McGraw-Hill.
Blommestein, H. et al. 1980. "Shopping Perceptions and Preferences: A Multidimensional Attractiveness Analysis of Consumer and Entrepreneurial Attitudes," *Economic Geography* 56:155-174.
Boswell, T. D. and T. C. Jones. 1978. "A Data Resource Base for Studying Persons of Spanish Origin in the United States," *The Professional Geographer* 49: 201-203.
Boswell, T. D. et al. 1982. "Liberty City — One Year After," Paper presented at the Annual Meeting of the Association of American Geographers, San Antonio, TX.
Bottomore, T. B. and M. Rubel (editors). 1956. *Karl Marx: Selected Writings in Sociology and Social Philosophy.* New York: McGraw Hill.
Bourne, L. S. 1976. "Urban Structure and Land Use Decisions," *Annals,* Association of American Geographers 66:531-547.
Bresser's 1984. *Bresser's Cross-Index Directory for Miami and Dade County.* Detroit, MI: Bresser's Cross-Index Directory Company.
Brooks, R. H. 1973. "Drought and Public Policy in Northeastern Brazil," *The Professional Geographer* 25:338-346.

Brown, M. A. and M. J. Broadway. 1981. "The Cognitive Maps of Adolescents: Confusion About Intertown Differences," *The Professional Geographer* 33:315-325.

Brunner, J. A. and G. A. Brunner. 1971. "Are Voluntary Unlisted Telephone Subscribers Really Different?," *Journal of Marketing Research* 8:121-124.

Burnett, K. P. 1980. "Spatial Constraints-Oriented Modelling: Empirical Analysis," *Urban Geography* 1: 153-166.

Burton, I. et al. 1970. *Suggestions for Comparative Field Observations on Natural Hazards.* Toronto: University of Toronto, Department of Geography, Natural Hazard Research Working Paper No. 16.

Cambridge Systematics, Inc. 1984. Personal communication.

Cantril, H. 1971. *Gauging Public Opinion.* Port Washington, NY: Kennikat Press.

Caruso, D. and R. Palm. 1973. "Social Space and Social Place," *The Professional Geographer* 25:221-225.

Chicago Area Transportation Study. 1962. *Final Report.* CATS.

Clark, D. and K. I. Unwin. 1980. *Information Services in Rural Areas: Prospects for Telecommunications Access.* Norwich, UK: Geobooks.

Clark, W. A. V. and J. O. Huff. 1980. "Sources of Spatial Variation in Residential Mobility Rates," *Urban Geography* 1: 202-214.

Cochran, W. G. 1963. *Sampling Techniques.* New York: John Wiley.

Converse, J. M. 1984. "Strong Arguments and Weak Evidence: The Open/Closed Controversy of the 1940's." *Public Opinion Quarterly* 48:267-282.

Couture, M. R. and T. C. Dooley. 1981. "Analyzing Traveler Attitudes to Resolve Intended and Actual Use of a New Transit Service," *Transportation Research Record* 794:27-32.

Cox, K. R. and J. J. McCarthy. 1980. "Neighborhood Activism in the American City: Behavioral Relationships and Evaluation," *Urban Geography* 1:22-38.

Cutter, S. 1982. "Residential Satisfaction and the Suburban Homeowner," *Urban Geography* 3:315-327.

Cybriwsky, R. A. 1978. "Social Aspects of Neighborhood Change," *Annals,* Association of American Geographers 68:17-33.

Daniel, W. W. 1979. *The Use of Random Digit Dialing in Telephone Surveys: An Annotated Bibliography.* Monticello, IL: Vance Bibliographies, P223.

Daniels, P. W. 1982. "An Exploratory Study of Office Location Behavior in Greater Seattle," *Urban Geography* 3:58-78.

Davis, D. R., Jr. and E. Cassetti. 1978. "Do Black Students Wish to Live in Integrated, Socially Homogenous Neighborhoods?", *Economic Geography* 54:197-209.

Dear, M. 1977. "Psychiatric Patients and the Inner City," *Annals,* Association of American Geographers 67:588-594.

Detroit Metropolitan Area Traffic Study. 1956. *Final Report.* DMATS.

Dillman, D. A. 1978. *Mail and Telephone Surveys, The Total Design Method.* New York: John Wiley.

Dixon, C. J. and B. E. Leach. 1978a. *Sampling Methods for Geographical Research.* Norwich, UK: Geo Abstracts.

Dixon, C. J. and B. E. Leach. 1978b. *Questionnaires and Interviews in Geographic Research.* Norwich, UK: Geo Abstracts.

Dixon, C. J. and B. E. Leach. 1984. *Survey Research in Underdeveloped Countries.* Norwich, UK: Geo Abstracts.

Domencich, T. A. and D. McFadden. 1975. *Urban Travel Demand.* Amsterdam: North Holland Publishing Company.

Donald, M. N. 1960. "Implications of Nonresponse for the Interpretation of Mail Questionnaire Data," *Public Opinion Quarterly* 24:99-114.

Downs, R. and D. Stea. 1973. *Image and Environment.* London: Edward Arnold.

Edwards, A. L. 1957. *Techniques of Attitude Scale Construction.* New York: Appleton-Century-Crofts.

Edwards, G. and D. Shaw. 1982. "The Use of Diary Techniques: The Rural Studies Case," *Geoforum* 13:251-256.

Feldman, E. J. 1981. *A Practical Guide to the Conduct of Field Research in the Social Sciences.* Boulder, CO: Westview Press.

Fowler, F. J., Jr. 1984. *Survey Research Methods.* Beverly Hills, CA: Sage Publications.

Frey, J. H. 1983. *Survey Research by Telephone.* Beverly Hills, CA: Sage Publications.

Friberg, J. C. 1974. *Survey Research and Field Techniques: A Bibliography for the Fieldworker.* Monticello, IL: Council of Planning Librarians, Exchange Bibliography 513.

Friedrichs, J. and H. Ludtke. 1975. *Participant Observation.* Lexington MA: Saon House Studies and Lexington Books.

Gauthier, H. L. and R. L. Mitchelson. 1981. "Attribute Importance and Mode Satisfaction in Travel Mode Choice Research," *Economic Geography* 57:348-361.

Gibson, L. J. and M. A. Worden. 1981. "Estimating the Economic Base Multiplier: A Test of Alternative Procedures," *Economic Geography* 57:146-159.

Glasser, G. J. and G. O. Metzger. 1972. "Random Digit Dialing as a Method of Telephone Sampling," *Journal of Marketing Research* 9:59-64.

Golledge, R. G. et al. 1973. "An Experimental Design for Recovering Cognitive Information About a City," Department of Geography, Ohio State University, Columbus, Ohio.

Gould, P. R. and R. White. 1974. *Mental Maps.* Baltimore, MD: Penguin Books.

Groves, R. M. and R. L. Kahn. 1979. *Surveys by Telephone.* New York: Academic Press.

Hagan, D. E. and C. M. Collier. 1983. "Must Respondent Selection Procedures for Telephone Surveys be Invasive?," *Public Opinion Quarterly* 47:547-556.

Haggett, P. et al. 1977. *Locational Methods.* New York: John Wiley.

Halperin, W. C. et al. 1983. "Exploring Entrepreneurial Cognitions of Retail Environments," *Economic Geography* 59:3-15.

Halvorson, P. L. 1973. "The Income Factor in the Journey to Work," *The Professional Geographer* 25:357-362.

Hanson, S. 1980. "The Importance of Multipurpose Journey to Work in Urban Travel Behavior," *Transportation* 9:229-248.

Hanson, S. and P. Hanson. 1981. "The Travel Activity Patterns of Urban Residents: Dimensions and Relationships to Sociodemographic Characteristics," *Economic Geography* 57:332-347.

Harvey, D. W. 1969. *Explanation in Geography.* New York: St. Martin's Press.

Harvey, M. E. et al. 1979. "Cognition of a Hazardous Environment: Reactions to Buffalo Airport Noise," *Economic Geography* 55:263-286.

Heggemeier, K. S. 1982. "Putting Surveys in Their Context," *NewsReport* 22:8-12.

Henderson, F. M. and M. P. Voiland, Jr. 1975. "Some Possible Effects of Energy Shortages on Residential Preferences," *The Professional Geographer* 27:323-326.

High School Geography Project. 1979. *Manufacturing and Agriculture, Unit 2.* New York: Macmillan.

Highsmith, R. M., Jr. 1962. "Suggestions for Improving Geographical Interview Techniques," *The Professional Geographer* 33:53-55.

Holsti, O. R. 1969. *Content Analysis for the Social Sciences and Humanities.* Reading, MA: Addison-Wesley.

Horvath, R. J. 1970. "On the Relevance of Participant Observation," *Antipode* 2:30-37.

Hursh-César, G. and P. Roy. (editors). 1976. *Third World Surveys: Survey Research in Developing Nations.* New Delhi: Macmillan.

Jacob, H. 1984. *Using Published Data, Errors and Remedies.* Beverly Hills, CA: Sage Publications.

Janelle, D. and M. Goodchild. 1983. "Diurnal Patterns of Social Group Distributions in a Canadian City," *Economic Geography* 59:403-425.

Johnson, J. H., Jr. and S. D. Brunn. 1980. "Residential Preference Patterns of Afro-American College Students," *The Professional Geographer* 32:37-42.

Jordan, T. G. and L. Rowntree. 1982. *The Human Mosaic, Third Edition.* New York: Harper and Row.

Joseph, A. E. and A. Poyner. 1982. "Interpreting Patterns of Public Safety Utilization in Rural Areas," *Economic Geography* 58:262-273.

Kahn, R. and C. F. Cannell. 1958. *Dynamics of Interviewing.* New York: John Wiley.

Kaiser Transit Group. 1982. *The 1980 On-Board Transit Survey, Methodology Report.* Miami, FL.

Kalton, G. 1983. *Introduction to Survey Sampling.* Beverly Hills, CA: Sage Publications.

King, L. J. and R. G. Golledge. 1978. *Cities, Space, and Behavior.* Englewood Cliffs, NJ: Prentice-Hall.

Kish, L. 1965. *Survey Sampling.* New York: John Wiley.

Kniffen, F. 1962. "The Tape Recorder in Field Research," *Annals,* Association of American Geographers 52:83.

Kocher, D. J. and T. J. Bell. 1977. "The Effects of Workplace-Residence Separation and Ride-Sharing on Employee Absenteeism: Survey and Pilot Study," *The Professional Geographer* 29:272-277.

Krippendorf, K. 1980. *Content Analysis, An Introduction to its Methodology.* Beverly Hills, CA: Sage Publications.

Kromm, D. E. 1973. "Response to Air Pollution in Ljubljana, Yugoslavia," *Annals,* Association of American Geographers 63: 208-217.

Kuzmyak, J. R. and S. Prensky. 1979. "Use of Travel Diaries in Collection of Travel Data on the Elderly and Handicapped," *Transportation Research Record* 701:36-38.

Leinbach, T. R. 1973. "Distance Information Flows and Modernization: Some Observations from West Malaysia," *The Professional Geographer* 25:7-11.

Lentnek, B. et al. 1978. "Commercial Factors in the Development of Urban Systems: A Mexican Case Study," *Economic Geography* 54-291-308.

Lentnek, B. et al. 1975. "Consumer Behavior in Different Environments," *Annals,* Association of American Geographers 65:538-564.

Ley, D. 1974. *The Black Inner City as Frontier Outpost: Images and Behavior of a Philadelphia Neighborhood.* Washington, D.C.: Association of American Geographers.

Lloyd, R. and D. Jennings. 1978. "Shopping Behavior and Income: Comparisons in an Urban Environment," *Economic Geography* 54:157-167.

Lloyd, R. and T. Steinke. 1977. "Visual and Statistical Comparison of Choropleth Maps," *Annals,* Association of American Geographers 67:429-436.

Lounsbury, J. F. and F. T. Aldrich. 1979. *Introduction to Geographic Field Methods and Techniques.* Columbus, OH: Charles E. Merrill.

Lyons, W. and R. F. Durant. 1980. "Interviewer Costs Associated with the Use of Random Digit Dialing in Large Area Samples," *Journal of Marketing* 44:65-69.

MacEachren, A. M. 1980. "Travel Time as the Basis of Cognitive Distance," *The Professional Geographer* 32:30-35.

Marble, D. F. et al. 1972. *Household Travel Behavior Study, Report No. 1, Field Operations and Questionnaires.* Evanston, IL: Northwestern University, The Transportation Center.

McConnell, J. E. 1979. "The Export Decision: An Empirical Study of Firm Behavior," *Economic Geography* 55:171-183.

McGrath, W. R. and C. Guinn. 1963. "Simulated Home Interview by Telephone," *Highway Research Record* 41:1-6.

McLafferty, D. and F. L. Hall. 1982. "The Use of Multinomial Logit Analysis to Model the Choice of Time to Travel," *Economic Geography* 58:236-246.

Memmott, F. W. 1963. "Home Interview Survey and Data Collection Procedure," *Highway Research Record* 41:7-12.

Mendenhall, W. et al. 1971. *Elementary Survey Sampling.* Belmont, CA: Duxbury Press.

Mercer, J. R. and D. A. Phillips. 1981. "Attitudes of Homeowners and the Decision to Rehabilitate Property," *Urban Geography* 2:216-236.

Meyer, J. W. 1981. "Migration to New Metropolitan Areas: Characteristics and Motives," *Urban Geography* 2:64-79.

Miami Herald. 1983. December 18, 1983:6E.

Miller, E. J. and M. E. O'Kelly. 1983. "Estimating Shopping Destination Choice Models from Travel Diary Data," *The Professional Geographer* 34:440-449.

Mitchelson, R. L. and H. L. Gauthier. 1981. "Gender-Related Differences in Travel Mode Psychophysics," *Urban Geography* 2:131-147.

Monroe, C. B. and P. L. Halvorson. 1980. "The Impact of Changes in Pricing Policy Upon Transit Riders of Varying Ages," *The Professional Geographer* 32:335-342.

Morrill, R. M. 1965. "The Negro Ghetto: Problems and Alternatives," *Geographical Review* 55:339-361.

NCHSR. 1977. *Advances in Health Survey Research Methods.* Washington, DC: United States Public Service.

Nie, N. H. et al. 1975. *Statistical Package for the Social Sciences.* New York: McGraw Hill.

Norback, C. 1980. *The Complete Book of American Surveys.* New York: Signet Books.

Norton, W. 1984. Historical Analysis in Geography. London, UK: Longman.

Nunnally, J. C. 1957. *Psychometric Methods.* New York: McGraw-Hill.

Ojo, G. J. A. 1973. "Journey to Agricultural Work in Yorubaland," *Annals,* Association of American Geographers 63:85-96.

O'Kelly, M. E. 1983. "Impacts of Multistop, Multipurpose Trips on Retail Distribution," *Urban Geography* 4:173-190.

Pacione, M. 1983. "The Temporal Stability of Perceived Neighborhood Areas in Glasgow," *The Professional Geographer* 34: 66-73.

Palm, R. 1973. "Factorial Ecology and the Community of Outlook," *Annals,* Association of American Geographers 63:341-346.

Peterson, R. A. 1984. "Asking the Age Questions: A Research Note," *Public Opinion Quarterly* 48:379-383.

Rand Corporation. 1966. *A Million Random Digits with 100,000 Normal Deviates.* New York: Free Press.

Rathje, W. L. and C. K. Ritenbaugh. (editors). 1984. "Household Refuse Analysis: Theory, Method, and Applications in the Social Sciences," *American Behavioral Scientist,* 28 (Special Issue).

Recker, W. W. and H. J. Schuler. 1982. "An Integrated Analysis of Complex Travel Behavior and Urban Form Indicators," *Urban Geography* 3:110-120.

Roseman, C. C. and P. L. Knight, III. 1975. "Residential Environment and Migration Behavior of Blacks," *The Professional Geographer* 27:160-165.

Roseman C. C. and J. D. Williams. 1980. "Metropolitan to Nonmetropolitan Migration: A Decision-Making Perspective," *Urban Geography* 1:283-294.

Rowles, G. D. 1978. *Prisoners of Space? Exploring the Geographic Experience of Older People.* Boulder, CO: Westview Press.

Rowley, G. 1984. Personal communication.

Rushton, G. 1969. "The Scaling of Locational Preferences," in K. Cox and R. G. Golledge (editors), *Problems of Spatial Behavior: A Symposium.* Evanston, IL: Department of Geography, Northwestern University.

Saarinen, T. 1976. *Environmental Planning: Perception and Behavior.* Boston, MA: Houghton Mifflin.

Sapp, M. and I. M. Sheskin. 1983. *Survey of Academic Computing at the University of Miami.* Coral Gables, FL: Office of Information Systems, University of Miami.

Schimpeler.Corradino Associates. 1980. *Final Report, Washtenaw County (Michigan) Transportation Needs Study.* Coral Gables, FL: Schimpeler.Corradino Associates.

Schimpeler.Corradino Associates. 1983. *Florida Statewide Transit Needs Assessment.* Coral Gables, FL: Schimpeler.Corradino Associates.

Schuler, H. J. 1979. "A Disaggregate Store-Choice Model of Spatial Decision-Making," *The Professional Geographer* 31:146-159.

Schuman, H. and S. Presser. 1978. "Question Wording as an Independent Variable in Survey Analysis," pp. 27-46 in Alwin 1978a.

Scott, A. J. 1983. "Industrial Organization and the Logic of Intra-Metropolitan Location, II: A Case Study of the Printed Circuits Industry in the Greater Los Angeles Region," *Economic Geography* 59:343-367.

Scott, A. J. 1984. "Industrial Organization and the Logic of Intra-Metropolitan Location, III: A Case Study of the Women's Dress Industry in the Greater Los Angeles Region," *Economic Geography* 60:3-27.

Seley, J. E. 1981. "Targeting Economic Development: An Examination of the Needs of Small Businesses," *Economic Geography* 57:34-51.

Shannon, G. W. et al. 1975. "A Method for Evaluating the Geographic Accessibility of Health Services," *The Professional Geographer* 27:30-36.

Sheskin, I. M. 1974. "The Social Trip Behavior of Suburban Residents," unpublished Master's thesis, Department of Geography, State University of New York at Buffalo.

Sheskin, I. M. 1982. *Methodology Report, Population Study of the Greater Miami Jewish Community.* Coral Gables, FL: University of Miami.

Sheskin, I. M. 1984. *Miami Review Readership Survey, Final Report.* Coral Gables, FL.

Sheskin, I. M. and H. A. Friedman. 1981. "Aging and Mobility in Dade County, Florida," *Florida Geographer* 15:17-24.

Sheskin, I. M. et al. 1981. "The Dade County On-Board Survey," *Transit Journal* 7:15-28.

Sheskin, I. M. and P. R. Stopher. 1982. "Pilot Testing of Alternative Administrative Procedures and Survey Forms," *Transportation Research Record* 886:8-22.

Sheskin, I. M. and R. Warburton. 1983. *University of Miami Travel and Parking Survey Final Report,* Coral Gables, FL: Office of Business Affairs, University of Miami.

Simon, J. L. 1978. *Basic Research Methods in Social Science.* New York: Random House.

Simpson-Housley, P. and F. A. Curtis. 1983. "Earthquake Occurrence, Experience and Appraisal in Wellington, New Zealand," *The Professional Geographer* 34:462-467.

Slonim, M. J. 1960. *Sampling.* New York: Simon and Schuster.

Smitt, B. and M. Flaherty. 1981. "Resident Attitudes Toward Exurban Development in a Rural Ontario Township," *The Professional Geographer* 33:103-112.

Spector, A. N. *et al.* **1976.** "Acquaintance Circles and Communication: An Exploration of Hypotheses," *The Professional Geographer* 28:267-276.

Sternstein, L. 1974. "Migration to and From Bangkok," *Annals,* Association of American Geographers 64:138-147.

Stimson, R. J. and E. S. Ampt. 1972. "Mailed Questionnaires and the Investigation of Spatial Behavior: The Problem of Respondent and Nonrespondent Differences," *Australian Geographer* 12:51-54.

Stoddard, R. H. 1982. *Field Techniques and Research Methods in Geography.* Dubuque, IA: Kendall/Hunt.

Stopher, P. R. 1982. "Small-Sample Home-Interview Travel Surveys: An Application and Suggested Modifications," *Transportation Research Record* 886:41-47.

Stopher, P. R. and A. Meyburg. 1975. *Urban Transportation Modelling and Planning.* Lexington, MA: Lexington Books.

Stopher, P. R. and I. M. Sheskin. 1982a. "Toward Improved Collection of a 24-hour Travel Record," *Transportation Research Record* 891:10-17.

Stopher, P. R. and I. M. Sheskin. 1982b. "Method for Determining and Reducing Nonresponse Bias," *Transportation Research Record* 886:35-41.

Stuart, A. 1968. *Basic Ideas of Scientific Sampling.* London, UK: Griffen.

Stutz, F. P. 1974. "Interactance Communities Vs. Named Communities," *The Professional Geographer* 26:401-411.

Stutz, F. P. and R. Butts. 1976. "Environmental Trade-Offs for Travel Behavior," *The Professional Geographer* 28:167-171.

Sudman, S. 1980. "Improving the Quality of Shopping Center Sampling," *Journal of Marketing Research* 17:423-431.

Sudman, S. and N. M. Bradburn. 1974. *Response Effects in Surveys: A Review and Synthesis.* Chicago, IL: Aldine.

Sudman, S. *et al.* **1978.** "Modest Expectations: The Effects of Interviewers' Prior Expectations on Responses," pp. 47-58 in Alwin 1978a.

Symanski, R. 1974. "Prostitution in Nevada," *Annals,* Association of American Geographers 64:357-377.

Talarchek, G. M. 1982. "Sequential Aspects of Residential Search and Selection," *Urban Geography* 3:34-57.

Taylor, P. 1977. *Quantitative Methods in Geography: An Introduction to Spatial Analysis.* Boston, MA: Houghton Mifflin.

Taylor, M. J. and R. J. W. Neville. 1981. "Whose Image of What? A Note on the Measurement of Industrialists' Attitudes and Images in Singapore," *The Professional Geographer* 33:335-340.

Thomas, R. N. 1971. "Survey Research Design: A Case of Rural Urban Mobility," pp. 421-427 in B. Lentnek *et al.* (editors), *Geographic Research on Latin America, Benchmark, 1970.* Muncie, IN: Ball State University, Proceedings of the Conference of Latin Americanist Geographers.

Timmermans, H. 1983. "Non-Compensatory Decision Rules and Consumer Spatial Choice Behavior," *The Professional Geographer* 34:449-455.

Tourism and Recreation Research Unit. 1983. *Recreation Site Survey Manual.* London, UK: E. & F. N. Spon.

Ulack, R. 1978. "The Role of Squatter Settlements," *Annals,* Association of American Geographers 68:535-550.

United Nations. 1971. *Methodology of Demographic Sample Surveys.* New York: United Nations.

United States Department of Commerce, Bureau of the Census. 1984. *Statistical Abstract of the United States.* Washington, DC: United States Government Printing Office.

Walker, R. 1976. "Social Survey Techniques: A Note on the Drop and Collect Method," *Area* 8:284-288.

Ward, J. 1975. "Skid Row as a Geographic Entity," *The Professional Geographer* 27:286-296.

Warwick, D. P. and C. A. Lininger. 1975. *The Sample Survey: Theory and Practice.* New York: McGraw-Hill.

Webb, E. J. et al. 1966. *Unobtrusive Measures: Nonreactive Research in the Social Sciences.* Chicago: Rand McNally.

Weiss, C. H. 1968. "Validity of Welfare Mothers' Interview Responses," *Public Opinion Quarterly* 32:622-633.

Wheeler, J. O. 1981. "Effects of Geographic Scale on Location Decisions in Manufacturing: The Atlanta Experience," *Economic Geography* 58: 134-145.

Wheeler, J. O. and R. N. Thomas. 1973. "Urban Transportation in Developing Economies, Work Trips in Tegucigalpa," *The Professional Geographer* 25:113-120.

Whyte, A. V. T. 1977. *Guidelines for Field Studies in Environmental Perception.* Paris: UNESCO.

Willcox, K. 1973. "A Trial of Methods for Collecting Household Morbidity Data," in J. Neil *et al.* (editors), *Genetics and the Epidemiology of Rare Diseases.* United States Public Health Service Publication No. 1163.

Winett, R. A. et al. 1979. "Effects of Self-Monitoring and Feedback on Residential Energy Consumption," *Journal of Applied Behavior Analysis* 12:173-184.

Yeates, M. 1974. *An Introduction to Quantitative Analysis in Human Geography.* New York: McGraw-Hill.

Young, M. and P. Willmott. 1973. *The Symmetrical Family.* London: Routledge and Kegan Paul.

Zelinsky, W. 1980. "America's Vernacular Regions," *Annals,* Association of American Geographers 70:1-16.

Zonn, L. E. 1980. "Information Flows in Black Residential Search Behavior," *The Professional Geographer* 32:43-50.

RESOURCE PUBLICATIONS

Resource Publications listed on the inside front cover should be ordered by author and title from the address below.

RESOURCE PAPERS

Resource Papers from the 1968-1974 series are available from the Association of American Geographers. A list of volumes in print will be sent upon request. Volumes in the 1975-78 Resource Papers series are also available:

1975 RESOURCE PAPERS

75/1-*LAND USE CONTROL: INTERFACE OF LAW AND GEOGRAPHY*, R. Platt, 1976, (ISBN 0-89291-109-3)

75/2-*THE OUTER CITY: GEOGRAPHICAL CONSEQUENCES OF THE URBANIZATION OF THE SUBURBS*, P. Muller, 1976, (ISBN 0-89291-114-X)

75/3-*TRIUMPH OR TRIAGE? THE WORLD FOOD PROBLEM IN GEOGRAPHICAL PER-SPECTIVE*, C.G. Knight, R.P. Wilcox, 1977, (ISBN 0-89291-115-8)

75/4-*MAPS, DISTORTION, AND MEANING*, M.S. Monmonier, 1977, (ISBN 0-89291-120-4)

1976 RESOURCE PAPERS

76/1-*THE GEOGRAPHY OF INTERNATIONAL TOURISM*, I.M. Matley, 1976, (ISBN 0-89291-110-7)

76/2-*SOCIAL ASPECTS OF INTERACTION AND TRANSPORTATION*, F.P. Stutz, 1977, (ISBN 0-89291-117-4)

76/3-*LANDSCAPE IN LITERATURE: A GEOGRAPHICAL ANALYSIS*, C. L. Salter, W. J. Lloyd, 1977, (ISBN 0-89291-118-2)

76/4-*GEOGRAPHY AND MENTAL HEALTH*, C.J. Smith, 1977, (ISBN 0-89291-119-0)

1977 RESOURCE PAPERS

77/1-*URBANIZATION AND ENVIRONMENTAL QUALITY*, T.R. Lakshmanan, L. Chatterjee, 1977, (ISBN 0-89291-122-0)

77/2-*CHANGING MIGRATION PATTERNS WITHIN THE UNITED STATES*, C.C. Roseman, 1977, (ISBN 0-89291-123-9)

77/3-*SPATIAL PERSPECTIVES ON SCHOOL DESEGREGATION*, J.D. Lord, 1977, (ISBN 0-89291-124-7)

77/4-*ENERGY: THE ULTIMATE RESOURCE*, E. Cook, 1978, (ISBN 0-89291-127-1)

1978 RESOURCE PAPERS

78/1-*THE GEOGRAPHY OF CRIME AND VIOLENCE*, D. Georges, 1978, (ISBN 0-89291-128-X)

78/2-*WATER RESOURCES FOR OUR CITIES*, D. Baumann, D. Dworkin, 1978, (ISBN 0-89291-130-1)

78/3-*ENVIRONMENTAL IMPACT STATEMENTS*, M.R. Greenberg, R. Anderson, G.W. Page, 1978, (ISBN 0-89291-131-X)

78/4-*SPATIAL ASPECTS OF AGING*, R. Wiseman, 1979, (ISBN 0-89291-133-6)

Requests for publication lists and prices, as well as standing orders, should be sent to:
Association of American Geographers
1710 Sixteenth Street, N.W.
Washington, D.C. 20009
(202-234-1450)

RESOURCE PUBLICATIONS IN GEOGRAPHY